T0219984

Haskell Quick Syntax Reference

A Pocket Guide to the Language, APIs, and Library

Stefania Loredana Nita

Marius Mihailescu

Apress®

Haskell Quick Syntax Reference: A Pocket Guide to the Language, APIs, and Library

Stefania Loredana Nita
Bucharest, Romania

Marius Mihailescu
Bucharest, Romania

ISBN-13 (pbk): 978-1-4842-4506-4
https://doi.org/10.1007/978-1-4842-4507-1

ISBN-13 (electronic): 978-1-4842-4507-1

Copyright © 2019 by Stefania Loredana Nita and Marius Mihailescu

This work is subject to copyright. All rights are reserved by the Publisher, whether the whole or part of the material is concerned, specifically the rights of translation, reprinting, reuse of illustrations, recitation, broadcasting, reproduction on microfilms or in any other physical way, and transmission or information storage and retrieval, electronic adaptation, computer software, or by similar or dissimilar methodology now known or hereafter developed.

Trademarked names, logos, and images may appear in this book. Rather than use a trademark symbol with every occurrence of a trademarked name, logo, or image we use the names, logos, and images only in an editorial fashion and to the benefit of the trademark owner, with no intention of infringement of the trademark.

The use in this publication of trade names, trademarks, service marks, and similar terms, even if they are not identified as such, is not to be taken as an expression of opinion as to whether or not they are subject to proprietary rights.

While the advice and information in this book are believed to be true and accurate at the date of publication, neither the authors nor the editors nor the publisher can accept any legal responsibility for any errors or omissions that may be made. The publisher makes no warranty, express or implied, with respect to the material contained herein.

Managing Director, Apress Media LLC: Welmoed Spahr
Acquisitions Editor: Steve Anglin
Development Editor: Matthew Moodie
Coordinating Editor: Mark Powers

Cover designed by eStudioCalamar

Cover image designed by Freepik (www.freepik.com)

Distributed to the book trade worldwide by Springer Science+Business Media New York, 233 Spring Street, 6th Floor, New York, NY 10013. Phone 1-800-SPRINGER, fax (201) 348-4505, e-mail orders-ny@springer-sbm.com, or visit www.springeronline.com. Apress Media, LLC is a California LLC and the sole member (owner) is Springer Science + Business Media Finance Inc (SSBM Finance Inc). SSBM Finance Inc is a **Delaware** corporation.

For information on translations, please e-mail editorial@apress.com; for reprint, paperback, or audio rights, please email bookpermissions@springernature.com.

Apress titles may be purchased in bulk for academic, corporate, or promotional use. eBook versions and licenses are also available for most titles. For more information, reference our Print and eBook Bulk Sales web page at www.apress.com/bulk-sales.

Any source code or other supplementary material referenced by the author in this book is available to readers on GitHub via the book's product page, located at www.apress.com/9781484245064. For more detailed information, please visit www.apress.com/source-code.

Printed on acid-free paper

Table of Contents

About the Authors

Stefania Loredana Nita has two B.Sc. degrees, one in mathematics (2013) and one in computer science (2016), from the Faculty of Mathematics and Computer Science at the University of Bucharest; she received her M.Sc. in software engineering (2016) from the Faculty of Mathematics and Computer Science at the University of Bucharest. She has worked as developer for an insurance company (Gothaer Insurance) and as a teacher of mathematics and computer science in private centers of education. Currently, she is a Ph.D. student in computer science in the Faculty of Mathematics and Computer Science at the University of Bucharest. She has been a teaching assistant at the same university and since 2015 has worked as a researcher and developer at the Institute for Computers in Bucharest, Romania. Her domains of interest are cryptography applied in cloud computing and big data, parallel computing and distributed systems, and software engineering.

Marius Mihailescu received his B.Sc. in science and information technology (2008) and his B.Eng. in computer engineering (2009) from the University of Southern Denmark; he has two M.Sc. degrees, one in software engineering (2010) from the University of Bucharest and the second one in information security technology (2011) from the Military Technical Academy. His Ph.D. is in computer science (2015) from the University of Bucharest, Romania, with a thesis on the security of biometrics authentication protocols. From 2005 to 2011 he worked as a software developer and researcher for different well-known companies (Softwin, NetBridge Investments, Declic) in Bucharest, Romania (working in software and web development, business analysis, parallel computing, cryptography researching, and distributed systems). From 2012 until 2015 he was an assistant in the informatics department at the University of Titu Maiorescu and in the computer science department of the University of Bucharest. Since 2015, he has been a lecturer at the University of South-East Lumina.

About the Technical Reviewer

Germán González-Morris is a polyglot software architect/engineer with 20+ years in the field, with knowledge in Java (EE), Spring, Haskell, C, Python, and JavaScript, among others. He works with web distributed applications. Germán loves math puzzles (including reading Knuth) and swimming. He has tech reviewed several books, including about an application container (WebLogic) and various programming languages (Haskell, TypeScript, WebAssembly, math for coders, and regexp). You can find more details on his blog (https://devwebcl.blogspot.com/) or on Twitter (@devwebcl).

Introduction

Haskell is a functional programming language, named after mathematician Haskell Brooks Curry, who made important contributions to mathematical logic that were used as the basis for functional programming. The first version, Haskell 1.0, was released in 1990, and the latest version is Haskell 2010, released in July 2010.

These are the main characteristics of Haskell:

- It is purely functional, which means that all functions written in Haskell are also functions in the mathematical sense. The variables are immutable; in other words, they cannot be changed by any expression. Haskell does not contain statements or instructions, just expressions that are evaluated.

- It is lazy, meaning the expressions are evaluated when it is really necessary. Combined with the purity of Haskell, you can create chains of functions, which improves performance.

- It is statically typed, which means that every expression has a type, established at compile time.

- It enables type inference through the unification of every type bidirectionally. This is beneficial, because you do not need to write specifically every type in Haskell, and if you need, you can define your own types.

- Haskell is concurrent because it works with effects.

- It has many open source packages.

From these features, you can identify some of the advantages of using Haskell, listed here:

- The quality of the code is high.

- You can work with abstract mathematical concepts.

- The type system is flexible.

- Errors are kept to a minimum.

- The syntax is optimized and well-designed.

- It brings a good performance due to concurrency.

In this book, we will start with simple topics and increase the level of complexity with each chapter. All the chapters contain examples to illustrate each specific subject covered.

Source Code

You can download this book's source code by navigating to `https://www.apress.com/us/book/9781484245064` and clicking the Download Source Code button.

Structure of the Book

This book contains 26 chapters. Each chapter has its own goal of presenting the most important aspects of one topic that need to be taken into consideration during the learning process. The book highlights all the necessary elements of functional programming in Haskell that you'll need to start developing programs with it.

The structure of the book is as follows:

- **Chapter 1: Functional Programming.** This chapter provides a short introduction to functional programming.

- **Chapter 2: Static Typing.** This chapter presents in a practical manner the elements necessary to understand how you can use static typing.

- **Chapter 3: GHC.** This chapter contains the basic ideas of the Glasgow Haskell Compiler, an open source native code compiler for the functional programming language. We show several examples of how you can use GHC.

- **Chapter 4: Types.** This chapter goes through all the details necessary to understand how types work and how you can use them.

- **Chapter 5: Tuples.** The discussion in this chapter focuses on how tuples are represented and implemented. We discuss their performance, and through different examples we show how they are best used.

- **Chapter 6: Lists.** The chapter demonstrates how lists can be implemented and their basic operations (finding/searching, adding, and deleting). In this chapter, you will learn how to modify a list or its elements and how to work with lists and I/O operations.

- **Chapter 7: Functions.** The chapter presents the main elements necessary to construct functions and how it is possible to work with them.

- **Chapter 8: Recursion.** The chapter shows the necessary elements for developing applications using recursion and the optimal way of doing it.

- **Chapter 9: List Comprehension.** The chapter shows how syntactic "sugar" such as list comprehension can be used in Haskell applications, which are designed as special applications.

- **Chapter 10: Classes.** The chapter discusses how classes are defined, gives examples, and shows how data can be structured and modeled with the help of classes.

- **Chapter 11: Pattern Matching.** This chapter covers some of the coolest syntactic constructs and how pattern matching can be applied.

- **Chapter 12: Monads.** This chapter covers monads and what they are. The examples shown demonstrate how programs, both generic and declarative, can be structured logically. The chapter shows how generic and declarative data types are transformed with the help of higher-order functions.

- **Chapter 13: Monad Transformers.** This chapter contains some examples that show the power of monad transformers, with the goal of building computations with effects.

- **Chapter 14: Parsec.** The chapter shows how you can use a parsec, an industrial-strength tool.

- **Chapter 15: Folds.** The chapter illustrates through examples the family of higher-order functions that process a data structure in some order and how the return value is constructed.

- **Chapter 16: Algorithms.** The chapter presents examples of algorithms and goes through different categories of algorithms such as currying, folds, design patterns, dynamic programming, and so on.

- **Chapter 17: Parsings.** The chapter demonstrates how you can use different methods of parsing.

- **Chapter 18: Parallelism and Concurrency.** In this chapter, we discuss about parallelism and how to speed up the code by making it run on multicore processors.

- **Chapter 19: Haskell Pipes.** This chapter includes practical examples of I/O operations and Haskell pipes.

- **Chapter 20: Lens.** The chapter discusses the `Control.Lens` package and how to use it to obtain maximum results with your Haskell program.

- **Chapter 21: Lazy Evaluation.** This chapter shows the main way to evaluate a Haskell program.

- **Chapter 22: Performance.** The chapter discusses techniques for increasing the performance of Haskell programs.

- **Chapter 23: Haskell Stack.** The chapter discusses the Stack tool and how you can use it in projects. It also discusses its dependencies.

- **Chapter 24: Yesod.** Yesod is a free and open source web framework developed for Haskell for productive development. The chapter gives a quick overview of Yesod and how to use it.

- **Chapter 25: Haskell Libraries.** The chapter shows how to use libraries and how to build new libraries.

- **Chapter 26: Cabal.** Cabal represents a common architecture for developing and building applications and libraries in Haskell. The chapter covers the packaging and destructions of software packages and how they can be used.

CHAPTER 1

Functional Programming

Functional programming represents a programming paradigm in which the computations are evaluated by mathematical functions. The paradigm avoids changing states and using mutable data.

Every function in Haskell is a function in the purest mathematical sense. I/O operations which generate side-effects are represented in a mathematical way.

This chapter introduces you to the advantages of functional programming and then compares it to object-oriented programming.

The Advantages of Functional Programming

There are several advantages of functional programming, listed here:

- **Free from side effects and bugs**: As mentioned, functional programming does not maintain state. Therefore, you will not have side effects. In conclusion, you can write code free from side effects and bugs.

- **Efficiency**: Functional programs consist of independent units that are able to run concurrently. This means the efficiency is higher.

© Stefania Loredana Nita and Marius Mihailescu 2019
S. L. Nita and M. Mihailescu, *Haskell Quick Syntax Reference*,
https://doi.org/10.1007/978-1-4842-4507-1_1

- **Lazy evaluation**: Functional programming supports lazy evaluation constructs such as lazy lists, lazy maps, and so on.

- **Nested functions**: Functional programming supports nested functions.

- **Parallel programming efficiency**: As mentioned, there is no mutable state, so there are no state-change issues.

Functional Programming vs. Object-Oriented Programming

Table 1-1 highlights the main differences between functional programming and object-oriented programming (OOP). This gives readers with experience in object-oriented programming a chance to understand what it means to move from an OOP paradigm to functional programming.

Table 1-1. *Functional vs. Object-Oriented Programming*

Functional Programming	OOP
Functional programming uses immutable data.	Object-oriented programming uses mutable data.
Functional programming is based on a declarative programming model.	Object-oriented programming is based on an imperative programming model.
The focus is on "what you are doing."	The focus is on "how you are doing."
Functional programming supports parallel programming.	Object-oriented programming is not suitable for parallel programming.
Functions do not have side effects.	The methods can produce serious side effects.

(continued)

Table 1-1. (*continued*)

Functional Programming	OOP
Flow control is based on using function invocation and function invocation with recursion.	Flow control is based on loops and conditional statements.
The collection data is iterated using a recursion concept.	Object-oriented programming uses a loop concept to iterate `Collection`. Data such as the `for-each` loop in C# or Java.
The execution order of statements is not important.	The execution order of statements is very important.
There is support for "abstraction over data" and "abstraction over behavior."	There is support only for "abstraction over data."

Summary

This chapter provided a short introduction to functional programming.

The chapters contained in this book cover the most important aspects and powerful elements (techniques, methods, and algorithms) that a beginner Haskell developer needs to get started with.

CHAPTER 2

Static Typing

In Haskell, the type system is quite detailed. From a theoretical point of view, it comes from typed Lambda calculus, introduced in 1930 by Alfonso Church,[1,2] where the *types* are automatically deducted from the way the objects are processed. Such programming languages are *statically typed*. More precisely, the processing is enforced by and based on the rules of a mathematical type system.

A *type system* is a set of rules used in a programming language to organize, build, and handle the types that will be accepted in the language. These rules focus on some important aspects, such as the following:

- Defining new types

- Associating types with different language constructs

- Focusing on type equivalence, which is important to determine when different types are the same

- Verifying type compatibility, which is useful to check whether the value of a specific type is correctly used in a given processing context

---

[1]Alfonso Church, https://en.wikipedia.org/wiki/Alonzo_Church
[2]Lambda calculus, https://en.wikipedia.org/wiki/Lambda_calculus

© Stefania Loredana Nita and Marius Mihailescu 2019
S. L. Nita and M. Mihailescu, *Haskell Quick Syntax Reference*,
https://doi.org/10.1007/978-1-4842-4507-1_2

- Deducing the type of the language when it is not declared, which can be done by applying rules for synthesizing the type of a construct from the types of its components

As you already know, Haskell is a fully static, scoped language, and the top-level variables have static scope. In other words, the whole program module will contain their definition. This provides *referential transparency* if no side effect will rise. This is an example of a top-level expression:

```
variable = expression
```

In this expression, `variable` has `expression` as its value.

Currying and Uncurrying

Currying represents the process of transforming a function that has multiple arguments in a tuple with the same type as the arguments into a function that will take just a single argument and will return another function that will accept further arguments, one by one.

The following expression:

```
g :: x -> (y  -> z)   -- The above expression can be written as
g :: x -> y -> z
```

represents the curried form of the following:

```
h :: (x, y) -> z
```

You can convert these two types in any direction with the help of the prelude functions `curry` and `uncurry`.

```
g = curry h
h = uncurry g
```

If you take a closer look, both forms are equally expressive. They will hold the following:

```
g x h = h (x,y)
```

In Haskell, all functions are considered curried. That is, all functions in Haskell are able to take just one argument. In the examples, you will see that this is mostly hidden in the notation.

For example, the first expression defined next will attach the variable plus to a curried function. A curried function is a functional closure. The second expression declared will add factorial to a recursive function, which will compute n!.

```
plus = \ a -> (\ b -> a + b)
plus :: Int -> Int -> Int
factorial = \ n -> if n==1 then 1 else n*factorial(n-1)
factorial :: Int -> Int
```

Observe that the expression produces two values: the value that the expression has and the type that the value has. The type of the function is called *signature*. For example, the value of plus is a function of a domain in which int represents the range. The type of the expression does not need to be explicitly declared; it is automatically deduced based on the types and the expression's components. Type deduction involves the verification of the type as a natural task or subtask. As a general conclusion, an n-ary curried function can be defined as \ p1 p2 ... pn -> expression. The function plus can be declared as follows:

```
plus = \ a b -> a + b
plus :: int -> int -> int
```

This is the short notation of the same thing:

```
plus a b = a + b
```

The functions can be used in infix and prefix forms. For example, the application (+) a b is equivalent to the more familiar a + b, and mod a b can be written a 'mod' b. (op) corresponds to the prefixed form of an infix binary operator, while 'op' corresponds to the infixed form of a prefix binary operator.

```
comp m n = (\ a -> m(n a))
comp :: (b -> c) -> (d -> e) -> d -> e
ff = (\ a -> a*a) `comp` (\ a -> a+a)
ff :: Integer -> Integer
Output:
ff 2
16
```

In the previous example, the types that can be found in the signature of the composition function comp are not defined as constants. They represent generic types, which are represented by type variables in the format *identifier*. As you can observe, the function declared, comp, is polymorphic. This means that the arguments the function can take are of any type that obey its signature.

Scoping Variables

The *scoping variables* represent an extension of Haskell's type system, which allows free type variables to be reused in the scope of a function. Consider the following function[3] as an example:

```
mkpair1 :: forall a b. a -> b -> (a,b)
mkpair1 aa bb = (ida aa, bb)
    where
```

---

[3]https://wiki.haskell.org/Scoped_type_variables

```
ida :: a -> a -- This refers to a in the function's type
signature
ida = id
```

```
mkpair2 :: forall a b. a -> b -> (a,b)
mkpair2 aa bb = (ida aa, bb)
    where
        ida :: b -> b -- Illegal, because refers to b in type
signature
        ida = id
```

```
mkpair3 :: a -> b -> (a,b)
mkpair3 aa bb = (ida aa, bb)
    where
        ida :: b -> b -- Legal, because b is now a free variable
        ida = id
```

It is better to avoid scoped type variables because they are not available in all compilers. A solution is available in Haskell 98.[4]

For example, the following can be interpreted as x 'asTypeOf' y and has the same value as x, but the deduced type says that x and y have the same type.

```
asTypeOf :: a -> a -> a
asTypeOf a b = a
```

Let's look at the following examples for the let declaration:

```
let {var₁ = expr₁; var₂ = expr₂; ..., varₙ = exprₙ) in expr
expr where {var₁ = expr₁; var₂ = expr₂; ..., varₙ=exprₙ}
```

The scope of var_i (where i = 1,n) represents the whole expression that contains the definition of var_i. The variable var_i is tied to the unevaluated

---

[4]https://www.haskell.org/onlinereport/

expression expr$_i$ (keep in mind that the evaluation in Haskell is lazy, as you will see in Chapter 16). The result of expr is represented by the let expression.

```
is_even = let {is_even n = n == 0 || n > 0 && is_odd(n-1);
is_odd n = n == 1 || n > 1 && is_even(n-1)}
in is_even
is_even:: Integer -> Bool

is_even' = is_even where
{is_even n = n == 0 || n > 0 && is_odd(n-1);
is_odd n = n == 1 || n > 1 && is_even(n-1)}
```

Types

As mentioned, Haskell is statically typed.

From an algebraic point of view, a type is a triple like T=<V, Op, Ax>. Here, V is the set of type values (the carrier set of the type), Op is defined as the set of type operators (including the signatures of the operators), and Ax represents the set of axioms describing the behavior and how the operators interact. To illustrate this, let's consider the following example:

```
type α list is
Op
        []: → α list                  // a list of the
                                         constructors
      : : α x α list →α             // [] represents the empty
                                        list, : is like cons
```

Ax

head: α list \ { [] } $\to \alpha$ // a list of the selectors

tail: α list \ { [] } $\to \alpha$ // head as carrier, tail is
                                              as a cdr

null: α list $\to$ bool   // a list of the predicates

L: α list, x: α

null [] = true   // testing if the list is
                      empty

null (x:L) = false

head(x : L) = x   // the head of the list

tail(x : L) = L   // the tail of the list

Summary

This chapter presented the most important aspects of static typing.
You learned about the following:

- Statically typing systems and how they are defined
  during a program/module

- How new types can be defined

- The rules behind the type system

- Indentation and its importance

- Types and how they are defined

References

1. A. Serrano Mena, *Beginning Haskell: A Project-Based Approach* (Apress, 2014).

2. Scoped type variables, `https://wiki.haskell.org/Scoped_type_variables`

3. Haskell 98 Language and Libraries, `https://www.haskell.org/onlinereport/`

CHAPTER 3

GHC

GHC stands for *Glasgow Haskell Compiler*, which is the native code compiler for Haskell. It is open source and can be downloaded from https://www.haskell.org/ghc/. It is also integrated into the Haskell Platform, available at https://www.haskell.org/downloads#platform. (We recommend you install the Haskell Platform because it contains the compiler, Cabal,[1] and other tools you'll use in this book.)

Introducing GHC

The main components of GHC are the interactive interpreter GHCi and the batch compiler. GHC contains many libraries, but you can also use other extensions, such as for concurrency, exceptions, the type system, quantifications, a foreign function interface, and so on. When a program is compiled, the default behavior of GHC is to compile quickly but not to optimize the generated code. Using the batch compiler with GHCi has some advantages: the library code is precompiled, and a user's program is compiled very quickly.

Let's look at how to use GHC in a terminal. In these examples, the operating system is Microsoft Windows 10, and we're using GHCi version 8.4.3. So, open a terminal and type the ghci command (if this command is not recognized in the terminal, then it needs to be added to the

---

[1]Cabal is a system that builds and packages Haskell libraries and programs.

© Stefania Loredana Nita and Marius Mihailescu 2019
S. L. Nita and M. Mihailescu, *Haskell Quick Syntax Reference*,
https://doi.org/10.1007/978-1-4842-4507-1_3

environment variables). Now, the terminal should display the version of GHCi followed on the next line by Prelude> (as in Figure 3-1).

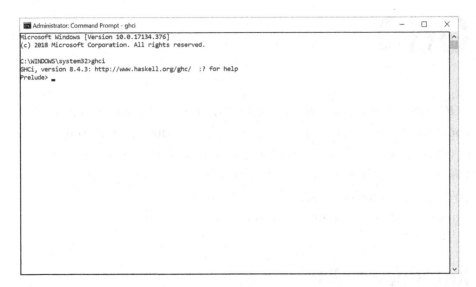

Figure 3-1. *Using GHCi in a terminal*

The GHCi represents GHC's interactive environment, so the *i* in the command ghci stands for *interactive*. Prelude is a standard module in Haskell (a module is a set of connected functions, types, or type classes), imported by default into all Haskell modules.

Let's print the string "Hello, World!" to the terminal. To do that, type putStrLn "Hello, World!" after Prelude>. putStrLn is a function in the Prelude standard module that prints the argument on the standard output device and adds a new line after the argument. (We will talk more about functions in the next chapters.) You'll get Hello, World!, as shown here:

```
Prelude> putStrLn "Hello, World!"
Hello, World!
```

Of course, real-world programs will be much more complex. It is more convenient to write them organized into files than to write them line by line at the terminal like this. The extension for Haskell files is .hs. So, in a text editor, write the following line:

```
main = putStrLn "This is my first Haskell program!"
```

and save it as first.hs. To compile the program, write the following line at the terminal:

```
ghc -o first first.hs
```

If you don't receive any error message, then the output should look like this:

```
[1 of 1] Compiling Main             ( first.hs, first.o )
Linking first.exe ...
```

If you receive the error <no location info>: error: can't find file: first.hs, make sure the path before the compiling command is the one that contains the first.hs file or put the file's full path in the compiling command. Here's an example:

```
ghc -o first D:\Haskell\FirstProgram\first.hs
```

The –o option tells the compiler to optimize compiler flags in order to generate faster code.

If you want to see more options or how a command works, type --help after the desired command.

To run the executable, type first.exe and then press Enter. Here's an example:

```
D:\Haskell\FirstProgram>first.exe
```

If you use Unix, just type `first` (without the extension). You will get the following:

```
This is my first Haskell program!
```

For Windows users, another way to use GHCi is to launch WinGHCi (Figure 3-2).

Figure 3-2. *A graphical user interface (GUI) for GHCi in Windows*

Examples

For the next examples, you will keep the standard prompt of `Prelude`. Note that Haskell is case sensitive, so a is different from A. You already used strings in the previous section, so let's look at some simple arithmetic operations.

```
Prelude> 10 + 5
15
Prelude> 100 - 20
80
Prelude> 14 * 88
1232
```

```
Prelude> 78 / 5
15.6
Prelude> (27 - 15) * 42
504
```

If you try 12 * -7, you will get an error, as shown here:

```
Prelude> 12 * -7
```

```
<interactive>:12:1: error:
    Precedence parsing error
        cannot mix '*' [infixl 7] and prefix '-' [infixl 6] in
the same infix expression
```

But if you try 12 * (-7), you will get the right answer.

```
Prelude> 12 * (-7)
-84
```

GHC contains predefined mathematical functions such as succ, min, max, div, and so on. Let's take a deep look at the div function. Its type is as follows:

```
Integral a => a -> a -> a
```

You should call the function like this:

```
Prelude> div 15 3
5
```

But Haskell permits you to use it in a more natural way.

```
Prelude> 15 'div' 3
5
```

Functions are the most important elements in Haskell; in the next example, you'll write a simple function that sums two numbers.

```
Prelude> mySum a b = a + b
```

17

In this definition, mySum is the name of the function, and a and b are the parameters. The part to the right of the equal sign is what the function does; in these examples, it adds the parameters. To call the function, you just type the name of the function and pass the arguments.

```
Prelude> mySum 7 8
15
```

So, there it is. In the next chapters, you will see that there are more types of functions, and you will write your own complex functions.

Summary

This chapter presented basic concepts about Haskell's compiler.

- You learned how to use GHC in two different ways.

- You compiled and ran some introductory examples.

References

1. Glasgow Haskell compiler, https://en.wikipedia.org/wiki/Glasgow_Haskell_Compiler

2. Haskell in five steps, https://wiki.haskell.org/Haskell_in_5_steps

3. GHC user's guide, https://downloads.haskell.org/~ghc/latest/docs/users_guide.pdf

4. P. Hudak, J. Hughes, S. Peyton Jones, and P. Wadler, "A history of Haskell: being lazy with class," in proceedings of the third ACM SIGPLAN conference on the history of programming languages, pp. 12–1 (ACM, 2007).

CHAPTER 4

Types

In Chapter 3, you saw that GHC contains predefined functions, and you worked a little with the `div` function. In that example, we subtly introduced *types*, without a comprehensive discussion. It is time to discuss them now. So, in this chapter, you'll learn about the main types in Haskell, how you can define your own types, and how the type system in Haskell works.

Basic Types in Haskell

The following are the main types in Haskell:

- `Char` represents a Unicode character.

- `Bool` represents a Boolean value. There are just two Boolean values: `True` and `False`.

- `Int` is an integer and represents a bounded type, which means it has a maximum value and a minimum value. These two values depend on the machine. For a 32-bit machine, the minimum value is `-2147483648`, and the maximum value is `2147483647`.

- The `Integer` type represents integer values also (very large integers, such as those used in cryptography), but its type is not bounded. Anyway, `Int` is more efficient.

© Stefania Loredana Nita and Marius Mihailescu 2019
S. L. Nita and M. Mihailescu, *Haskell Quick Syntax Reference*,
https://doi.org/10.1007/978-1-4842-4507-1_4

- Float represents values of floating-point numbers with single precision.

- Double represents values of floating-point numbers with double precision.

Next, let's take a look at the following examples:

```
Prelude> :t 5
5 :: Num p => p
Prelude> :t 'a'
'a' :: Char
Prelude> :t "abc"
"abc" :: [Char]
Prelude> :t False
False :: Bool
Prelude> :t 3 < 2
3 < 2 :: Bool
Prelude> :t 3.44
3.44 :: Fractional p => p
Prelude> :t pi
pi :: Floating a => a
```

In this code, the :t command shows the type of the expression that follows the command, the :: sign in the result means "has type of," and the words between the :: sign and the type is called the *type signature*.

You know that 5 is an integer, but you get Num here. Also, you know that 3.44 and pi are double values, and you get Fractional and Floating, respectively. The reason for this is that Num, Fractional, and Floating are *classes* in Haskell. For the moment, it is enough to know that Num includes Int, Integer, Double, and Float types; Floating includes Double and Float types; and Fractional includes Double and Float. You will learn more about classes in Chapter 10, and you will see the differences between them.

Another interesting part of the code is [Char]. The square brackets—[ ]—means that the evaluated expression is a *list*. In this example's case, it is a list of Chars, which is equivalent to String.

Next, let's *bind* locally a variable with a value. You can do this in two ways: using the let keyword or simply using the name of the variable followed by an equal sign, followed by the value, as shown here:

```
Prelude> let x = 1
Prelude> a = 'x'
Prelude> :t x
x :: Num p => p
Prelude> :t a
a :: Char
Prelude> :info x
x :: Num p => p  -- Defined at <interactive>:49:5
Prelude> :info a
a :: Char    -- Defined at <interactive>:50:1
```

In this code, :info is similar to :t, except that it displays some additional information.

Some useful structures are *lists* and *tuples*. A list (represented with squared brackets, [ ]) is a collection of elements of the same type (as [Char] showed earlier), while a tuple (represented with parentheses, ( )) is a collection of elements of different types. A tuple can be a list, but the reverse is not true. You will learn more about lists and tuples in Chapter 6, but for the moment, let's see how they look:

```
Prelude> [1, 2, 3]
[1,2,3]
Prelude> :t ['a', 'b']
['a', 'b'] :: [Char]
Prelude> [1, 'x']
```

```
<interactive>:97:2: error:
    • No instance for (Num Char) arising from the literal '1'
    • In the expression: 1
      In the expression: [1, 'x']
      In an equation for 'it': it = [1, 'x']
Prelude> (1, 'x')
(1,'x')
Prelude> :t (1, 'x')
(1, 'x') :: Num a => (a, Char)
Prelude> (1,2,3,4,5)
(1,2,3,4,5)
```

Defining Your Own Types

Suppose you want to create a structure that simulates a date. You need three integer values corresponding to the day, the month, and the year. You can define it using the data keyword, as shown here:

```
Prelude> data DateInfo = Date Int Int Int
Prelude> myDate = Date 1 10 2018
Prelude> :t myDate
myDate :: DateInfo
Prelude> :info DateInfo
data DateInfo = Date Int Int Int -- Defined at
<interactive>:101:1
```

DateInfo is the name of your new type, and it is called a *type constructor*, which is used to refer to the type. The Date after the equal sign is the *value constructor* (or *data constructor*), which is used to create values of DateInfo type. The three Ints after Date are components of the type. Note that the name of the type constructor and the name of the value constructor begin with capital letters. If you want to print myDate, you would get an error, because the print function does not have an argument of type DateInfo. To print myDate, you need to modify the definition of DateInfo, adding deriving (Show) to the end of definition (we will talk in detail about deriving in Chapter 10). The procedure is shown here:

```
Prelude> print myDate

<interactive>:34:1: error:
    • No instance for (Show DateInfo) arising from a use of
'print'

             -- Defined at <interactive>:27:44
    • In the expression: print myDate
      In an equation for 'it': it = print myDate
Prelude> data DateInfo = Date Int Int Int deriving (Show)
Prelude> myDate = Date 1 10 2018
Prelude> print myDate
Date 1 10 2018
```

The comparison between two dates will give a similar error. You need to add Eq after Show, which tells the compiler that you allow comparison between two dates.

```
Prelude> data DateInfo = Date Int Int Int deriving (Show, Eq)
Prelude> myDate1 = Date 1 10 2018
Prelude> myDate2 = Date 15 10 2018
Prelude> myDate3 = Date 1 10 2018
Prelude> myDate2 == myDate1
False
Prelude> myDate3 == myDate1
True
```

Synonyms

Let's define another type, called StudentInfo. A student is described by name, birth date, and specialization, so StudentInfo looks like this:

```
Prelude> data StudentInfo = Student String DateInfo String
deriving (Show, Eq)
```

It is a little difficult to distinguish in this definition which is the student's name and which is the student's specialization. But you can "rename" the basic types, as shown here:

```
Prelude> type Birthdate = DateInfo
Prelude> type Name = String
Prelude> type Specialization = String
```

So, StudentInfo becomes as follows:

```
Prelude> data StudentInfo = Student Name Birthdate
Specialization deriving (Show, Eq)
Prelude> student = Student "Alice Brown" (Date 21 8 1992)
"Computer Science"
Prelude> :t student
student :: StudentInfo
```

This kind of structure is called a *product type*, and it represents a tuple or a constructor with at least two arguments.

Structures and Enumerations

StudentInfo in this example would be considered a struct in the C/C++ programming languages.

If you want a structure that enumerates the elements, you can do the following:

```
Prelude> data Color = Red | Green | Blue | Yellow | Purple |
Orange deriving (Show, Eq)
```

In this example, Color is a sum type and represents a type that can have multiple possible forms.

Now, let's say you want to describe the people in a faculty. These can be teachers, defined by name and subject, or students, defined by name, birth date, and specialization. You can write this as follows:

```
Prelude> data FacultyPerson = Teacher String String | Student
Name DateInfo Specialization deriving (Show, Eq)
Prelude> teacher = Teacher "Emily Brian" "Functional
programming"
Prelude> student = Student "James Lee" (Date 23 4 1990)
"Computer Science"
Prelude> print teacher
Teacher "Emily Brian" "Functional programming"
```

Records

Let's suppose you want to add more information in your StudentInfo type (for simplicity, you will use just Student). The student will be described by first name, last name, birth date, specialization, study year, and average grade. The Student type looks like this:

```
Prelude> data Student = Student String String DateInfo String
Int Float deriving (Show, Eq)
```

Note   The type constructor now has the same name as the data constructor. Haskell allows you to do this.

Another way other than type to make it more intuitive is to make it a record, as shown here:

```
Prelude> :{
Prelude| data Student = Student { firstName :: String
Prelude|                        , lastName :: String
Prelude|                        , birthDate :: DateInfo
Prelude|                        , specialization :: String
Prelude|                        , studyYear :: Int
Prelude|                        , averageGrade :: Float
Prelude|                        } deriving (Show, Eq)
Prelude| :}
Prelude> student = Student "Emily" "Brian" (Date 23 4 1990)
"Computer Science" 2 9.14
Prelude> firstName student
"Emily"
Prelude> averageGrade student
9.14
Prelude> :t averageGrade
averageGrade :: Student -> Float
```

In this piece of code, you can see :{ and :}. This means you are writing a command on multiple lines. You can use this just in GHCi, not in .hs files. Using record, you can easily access a field of the structure just by typing the name of the field followed by the name of the variable.

Type System

In Haskell, the system has the following types:

- **Strong type**: A *strong type system* ensures that the program will contain errors resulting from wrong expressions. An expression that meets all the conditions of a language is called *well-typed*; otherwise, it is *ill-typed* and will lead to a *type error*. In Haskell, strong typing does not allow automatic conversions. So, if a function has a Double argument but the user provides an Int parameter, then an error will occur. Of course, the user can explicitly convert the Int value to a Double value using the predefined conversion functions and everything will be fine.

- **Static type**: In a *static type system*, the types of all values and expressions are known by the compiler at compile type, before executing the program. If something is wrong with the types of an expression, then the compiler will tell you, as in the example of lists. Combining strong and static types will avoid runtime errors.

- **Inference type**: In an *inference type system*, the system recognizes the type of almost all expressions in a program. Of course, the user can define explicitly any variable, providing its type, but this is optional.

Summary

In this chapter, you learned the following:

- Which are the basic types in Haskell

- How to define your own types

- How the type system works in Haskell

References

1. M. Lipovaca, *Learn You a Haskell for Great Good! A Beginner's Guide* (No Starch Press, 2011)

2. C. McBride, "Faking It: Simulating Dependent Types in Haskell," *Journal of Functional Programming*, 12(4–5), 375–392 (2002)

3. N. Vazou, E. L. Seidel, R. Jhala, D. Vytiniotis, and S Peyton-Jones, S. "Refinement Types for Haskell" in ACM SIGPLAN Notices, vol. 49, no. 9, pp. 269–282 (ACM, 2014)

4. J. Hughes, "Restricted Data Types in Haskell," *Haskell Workshop*, vol. 99 (1999)

5. B. O'Sullivan, J. Goerzen, and D. B. Stewart, *Real World Haskell: Code You Can Believe In* (O'Reilly Media, 2008)

6. Values, types, and other goodies, `https://www.haskell.org/tutorial/goodies.html`

7. Type, `https://wiki.haskell.org/Type`

CHAPTER 5

Tuples

Sometimes in your applications you'll need to group elements. In this chapter, you will learn how to do this using tuples and what the main predefined functions for tuples are.

Well, tuples are simple. They are a group of elements with different types. Tuples are *immutable*, which means they have a fixed number of elements. They are useful when you know in advance how many values you need to store. For example, you can use them when you want to store the dimensions of a rectangle or store the details of a student in the example from Chapter 4 (but it would be more difficult to read and follow the logic in the code).

Writing Tuples

Tuples are written between regular parentheses, and the elements are delimited by commas.

```
Prelude> ("first", "second", "third")
("first","second","third")
Prelude> :t ("first", "second", "third")
("first", "second", "third") :: ([Char], [Char], [Char])
Prelude> (1, "apple", pi, 7.2)
(1,"apple",3.141592653589793,7.2)
Prelude> :t (1, "apple", pi, 7.2)
(1, "apple", pi, 7.2)
  :: (Floating c, Fractional d, Num a) => (a, [Char], c, d)
```

© Stefania Loredana Nita and Marius Mihailescu 2019
S. L. Nita and M. Mihailescu, *Haskell Quick Syntax Reference,*
https://doi.org/10.1007/978-1-4842-4507-1_5

```
Prelude> ("True", 2)
("True",2)
Prelude> :t ("True", 2)
("True", 2) :: Num b => ([Char], b)
Prelude> (True, 2)
(True,2)
Prelude> :t (True, 2)
(True, 2) :: Num b => (Bool, b)
```

In the first example, you have three elements. In the second example, you have four elements, and in the last two examples, you have two elements each. Note that "True" is different from True. The first one, in quotation marks, is a String (or [Char]) value, while the second one is a Bool value.

You call a tuple with two elements a *pair* and a tuple with three elements a *triple*.

Actually, a tuple with more than three elements is not so common. Tuples are useful when you need to return more values from a function. In Haskell, when you want to return more values from a function, you need to wrap them in a single data structure with a single purpose, namely, a tuple.

Note that a tuple can have another tuple as an element, as shown here:

```
Prelude> (5, 'a', (2.3, False, "abc", 4))
(5,'a',(2.3,False,"abc",4))
Prelude> :t (5, 'a', (2.3, False, "abc", 4))
(5, 'a', (2.3, False, "abc", 4))
  :: (Fractional a1, Num a2, Num d) =>
     (a2, Char, (a1, Bool, [Char], d))
```

Predefined Functions for Pairs

You have seen that a particular type of tuple is a pair. Pairs are more widely used than other tuples. Let's suppose you want to declare a point in the Cartesian space, giving the x-coordinate and y-coordinate.

```
Prelude> (3.2, 5.7)
(3.2,5.7)
Prelude> let point = (3.2, 5.7)
```

If you need the x-coordinate of point, Haskell has a predefined function for you.

```
Prelude> fst point
3.2
```

The fst function returns the first value in the pair. Similarly, the snd function returns the second value in the pair.

```
Prelude> snd point
5.7
```

Another predefined function is swap. It belongs to the Data.Tuple module, and to use it, you first need to import this module. Then, you need to declare again the point variable, because it does not exist in the new scope. Finally, call the swap function.

```
Prelude> import Data.Tuple
Prelude Data.Tuple> let point = (3.2, 5.7)
Prelude Data.Tuple> swap point
(5.7,3.2)
```

The result of swap is a pair with switched elements. For the moment, don't worry about modules and importing. You will learn more about modules in Chapter 7. Note that Data.Tuple contains two more functions: curry and uncurry (you will learn how they work in Chapter 7).

Using the predefined functions over pairs, let's compute the distance between two points, A and B, in the Cartesian space. The distance is given by the following formula:

$$d = \sqrt{(x_A - x_B)^2 + (y_A - y_B)^2}$$

This is easy, because you already have a function that computes the square root of a number (sqrt) and a function that raises a number to another number (the power sign, ^). Therefore, the distance would look like this:

```
Prelude> let pointA = (2.4, 6)
Prelude> let pointB = (-7, 3.5)
Prelude> d = sqrt ((fst pointA - fst pointB)^2 + (snd pointA -
snd pointB)^2)
Prelude> d
9.726767191621274
```

Summary

In this chapter, you learned what tuples are and when they are useful. In addition, you worked with predefined functions over tuples.

Also in this chapter, you looked at modules and saw a preview of how they can be imported.

References

1. C. Hall and J. O'Donnell, "Introduction to Haskell" in *Discrete Mathematics Using a Computer*, pp. 1–33 (Springer, 2000)

2. A. S. Mena, *Beginning Haskell: A Project-Based Approach* (Apress, 2014)

3. Data.Tuple, http://hackage.haskell.org/ package/base-4.12.0.0/docs/Data-Tuple.html

CHAPTER 6

Lists

In this chapter, you'll learn about lists and why are they so useful. You will learn what a list is, which basic functions there are for lists, which operations are faster, and in which context you might use lists.

Basic Functions on Lists

A list is a similar data structure to a tuple, but lists can be used in more scenarios than tuples. Lists are pretty self-explanatory, but you need to know that they are *homogenous* data structures, which means the elements are of the same type. You represent lists using square brackets, [ ]. Here are some examples:

```
Prelude> [1, 2, 3]
[1,2,3]
Prelude> ['a', 'x']
"ax"
Prelude> [True, False, False, True]
[True,False,False,True]
Prelude> :t [True, False, False, True]
[True, False, False, True] :: [Bool]
Prelude> let numList = [3, 1, 0.5]
Prelude> :t numList
numList :: Fractional a => [a]
Prelude> ['a', 5]
```

© Stefania Loredana Nita and Marius Mihailescu 2019
S. L. Nita and M. Mihailescu, *Haskell Quick Syntax Reference*,
https://doi.org/10.1007/978-1-4842-4507-1_6

```
<interactive>:13:7: error:
    • No instance for (Num Char) arising from the literal '5'
    • In the expression: 5
      In the expression: ['a', 5]
      In an equation for 'it': it = ['a', 5]
```

The first example is intuitive. In the next example, observe that Haskell has represented the list `['a', 'x']` as `String` "ax" because, as you learned, a `String` is actually a list of characters. You can find out the type of elements using the `:t` command, and you can give a name to your list using the `let` keyword. In the previous example, observe that you cannot create a list with elements of different types. Also, note that all the numbers in numList are represented as `Fractional`.

Note that `[]` represents an empty list (containing no elements), as shown here:

```
Prelude> []
[]
```

The `[]`, `[[]]`, and `[[], []]` examples are different. The first is an empty list, the next is a list with one element (an empty list), and the last is a list with two elements (two empty lists). You can check whether a list is null (i.e., an empty list) using the `null` function.

```
Prelude> null []
True
Prelude> null [3.2]
False
```

A list can have two parts, depending the perspective (you will see in a moment why we used can here): the *head* and the *tail*. The head is the first element of the list, and the tail is made up of the remaining elements. There are two function for this.

```
Prelude> head [0,1,2,3,4,5]
0
Prelude> tail [0,1,2,3,4,5,6]
[1,2,3,4,5,6]
```

Somewhat opposite to head and tail are the last and init functions. The last function returns the last element of the list, while the init function returns the entire list of elements *except* the last one.

```
Prelude> last [1,2,3,4,5,6,7]
7
Prelude> init [1,2,3,4,5,6,7]
[1,2,3,4,5,6]
```

You can add more elements to a list in two ways, as shown here:

```
Prelude> [1,2,3] ++ [4,5]
[1,2,3,4,5]
Prelude> "Haskell" ++ " " ++ "programming"
"Haskell programming"
Prelude> 0 : [1,2]
[0,1,2]
Prelude> [0] ++ [1,2]
[0,1,2]
```

You can use : or ++ operators. What is the difference? You use : when you want to add a new head to the list, so on the left side of : is an element and on its right side is a list. You use ++ when you want to *concatenate* two lists, so on both sides of the ++ operator you write lists. This fact is emphasized in the previous example, where, to use ++, we wrote [0], namely, a list with just one element. The list [5, 6, 7] is actually the condensed representation of the operations 5:6:7:[].

To determine the length of a list, you use the length function.

```
Prelude> length [1,2,3,4,5,6,7,8,9,10]
10
Prelude> length []
0
-
```

If you need an element in a certain position, you use the !! operator (note that the first index of a list is 0).

```
Prelude> "Haskell programming" !! 10
'o'
Prelude> let listOfLists = [[1,2,3], [0], [-5, 3, 8]]
Prelude> length listOfLists
3
Prelude> listOfLists !! 2
[-5,3,8]
```

You can compare the elements of two lists. The comparison begins with the first elements, followed by a comparison of the second elements, and so on.

```
Prelude> [1,2,3] < [4,5]
True
Prelude> [1,2,3] < [4,1,2]
True
```

If you have many elements in a list, it is difficult to write them all. For example, what if you need all numbers from 1 to 100 or the letters from a to z or all odd numbers? How do you write them? Well, it's simple in Haskell. Haskell provides you with the following way:

```
Prelude> [1..20]
[1,2,3,4,5,6,7,8,9,10,11,12,13,14,15,16,17,18,19,20]
Prelude> ['a'..'z']
```

```
"abcdefghijklmnopqrstuvwxyz"
Prelude> [1,3..20]
[1,3,5,7,9,11,13,15,17,19]
Prelude> [14,18..30]
[14,18,22,26,30]
```

Note that if you want to define a rule, you need to provide the first two elements of the list.

You can even work with infinite lists, for example multiples of 5: [5,10..]. Be careful when you work with infinite lists.

Other Functions

The following are other useful functions for lists:

- reverse reverses the list.

- take takes from the list a certain number of elements (resulting in another list, but without doing any changes in the elements because in Haskell expressions are immutable), beginning with the first element.

- drop deletes from the list a certain number of elements, beginning with the first element.

- maximum extracts the maximum element of the list.

- minimum extracts the minimum element of the list.

- sum sums the elements of the list.

- product computes the product of elements in the list.

- elem checks whether an item is an element of the list.

- splitAt splits a list into two lists at a certain position.

- cycle creates an infinite list by replicating a given list.

- repeat creates an infinite list by repeating a given element.

Here are some examples:

```
Prelude> reverse []
[]
Prelude> reverse [1,2,3]
[3,2,1]
Prelude> take 5 [1,2,3,4,5,6,7]
[1,2,3,4,5]
Prelude> take 5 [1,2,3,4]
[1,2,3,4]
Prelude> drop 5 [1,2,3,4]
[]
Prelude> drop 5 [1,2,3,4,5,6,7]
[6,7]
Prelude> maximum [10, 3, 6, 132, 5]
132
Prelude> minimum [10, 3, 6, 132, 5]
3
Prelude> minimum []
*** Exception: Prelude.minimum: empty list
Prelude> minimum [True, False]
False
Prelude> minimum ['a', 'b']
'a'
Prelude> sum [10, 3, 6, 132, 5]
156
Prelude> sum []
0
Prelude> sum [True, False]
```

```
<interactive>:50:1: error:
    • No instance for (Num Bool) arising from a use of 'sum'
    • In the expression: sum [True, False]
      In an equation for 'it': it = sum [True, False]
Prelude> sum ['a', 'b']

<interactive>:51:1: error:
    • No instance for (Num Char) arising from a use of 'sum'
    • In the expression: sum ['a', 'b']
      In an equation for 'it': it = sum ['a', 'b']
Prelude> product [10, 3, 6, 132, 5]
118800
Prelude>  elem 3 [10, 3, 6, 132, 5]
True
Prelude>  elem True [10, 3, 6, 132, 5]

<interactive>:54:13: error:
    • No instance for (Num Bool) arising from the literal '10'
    • In the expression: 10
      In the second argument of 'elem', namely '[10, 3, 6, 132,
      ....]'
      In the expression: elem True [10, 3, 6, 132, ....]
Prelude>  elem 0 [10, 3, 6, 132, 5]
False
Prelude> splitAt 3 [10, 3, 6, 132, 5]
([10,3,6],[132,5])
Prelude> take 15 (cycle [1,2,3,4])
[1,2,3,4,1,2,3,4,1,2,3,4,1,2,3]
Prelude> take 15 (repeat 2)
[2,2,2,2,2,2,2,2,2,2,2,2,2,2,2]
```

The operations that are always fast are the ones appending an element (i.e., the : operator), head function, or tail function. The functions that imply working with the nth element of a list work pretty well, too, but they become slower as n becomes larger. Intuitively, the slowest functions are those that process an entire list, and they become even slower when the length of the list increases.

Summary

In this chapter, you learned the following:

- What lists are

- What the basic functions for a list are

- That you can represent a list in different ways

- That you can work with infinite lists

- Which operations are faster and which are slower

References

1. G. Hutton, *Programming in Haskell* (Cambridge University Press, 2016)

2. D. Coutts, D. Stewart, and R. Leshchinskiy, "Rewriting Haskell Strings." International Symposium on Practical Aspects of Declarative Languages (Springer, 2007)

3. How to work on lists, https://wiki.haskell.org/How_to_work_on_lists

4. Data.List, http://hackage.haskell.org/package/base-4.12.0.0/docs/Data-List.html

CHAPTER 7

Functions

In the previous chapters, you mainly worked with predefined functions from different libraries in Haskell. Now, it's time to write your own functions. In this chapter, you will learn about pattern matching, guards, clauses, higher-order functions, and lambda expressions used in functions.

Let's start with a simple function that adds two numbers.

```
add :: Integer -> Integer -> Integer
add x y =  x + y

main = do
    putStrLn "Adding two numbers:"
    print(add 3 7)
```

On the first line is the function declaration, which tells you the type of inputs and outputs, and on the second line is the function definition. As in the other programming languages, Haskell begins to compile the code from the main function. The result is as follows:

```
Adding two numbers:
10
```

Haskell Files

Haskell files can have an .hs or .lhs extension. The .lhs files have a literate format; in other words, they include commentary, and the lines that begin with > are considered part of the code. To write a Haskell file,

© Stefania Loredana Nita and Marius Mihailescu 2019
S. L. Nita and M. Mihailescu, *Haskell Quick Syntax Reference*,
https://doi.org/10.1007/978-1-4842-4507-1_7

open a text editor and place your code in it; then save the file with an `.hs` extension. To load a file into GHCi, first change the current directory to the directory that contains the `.hs` file using the `:cd` command. So, if you save your Haskell files in the path `C:\Haskell`; you can proceed as follows:

```
Prelude> :cd C:\Haskell
```

In `C:\Haskell`, let's say you have the file `Main.hs`. You can load this file into GHCi using the `:load` or `:l` command.

```
Prelude> :l Main.hs
[1 of 1] Compiling Main                 ( Main.hs, interpreted )
Ok, one module loaded.
*Main>
```

Note that `Prelude>` changes to `*Main>`.

GHC finds the file that contains the module `M` by looking at the name of the file. So, module `M` must be in the `M.hs` file. This rule is broken when the files are loaded using the `:load` command. In this case, the module name can be different from the file name, but the file must contain that module. If a source file is modified, it needs to be saved and then reloaded using `:reload`, as shown here:

```
 *Main> :reload
[1 of 1] Compiling Main                 ( Main.hs, interpreted )
Ok, one module loaded.
```

To import external modules, such as modules from libraries (as you will see in the following chapters), you use the `import` command.

```
Prelude> import Data.Maybe
Prelude Data.Maybe>
```

Of course, modules can be imported in other modules (for example, in source files) using the following syntax:

```
module Module_Name where
    import Module1
    import Module2
```
...

The command **import** has some variations that allow you to import some parts of a module, give a module an alias, and so on.

As with many programming languages, Haskell has a function called main, which specifies that an I/O action will be performed. Open the Main. hs file and type the following:

```
main = putStrLn "Learning about Haskell main function"
```

Then save it, load it, and run it.

```
Prelude> :l Main.hs
[1 of 1] Compiling Main                ( Main.hs, interpreted )
Ok, one module loaded.
*Main> main
Learning about Haskell main function
```

To write more I/O actions in main, you use the do block. Open the Main.hs file again and type the following:

```
main = do
    putStrLn "Are you enjoying Haskell?"
    answer <- getLine
    putStrLn ("You answered: " ++ answer)
```

Save the file and reload it. Then run it again, as shown here:

```
*Main> :reload
*Main> main
Are you enjoying Haskell?
sure
You answered: sure
```

Pattern Matching

Pattern matching means the program checks whether some data matches a certain pattern and then acts accordingly. A function can have different bodies for different patterns. Pattern matching can be applied on any data type. Check out this example:

```
day :: (Integral a) => a -> String
day 1 = "Monday"
day 2 = "Tuesday"
day 3 = "Wednesday"
day 4 = "Thursday"
day 5 = "Friday"
day 6 = "Saturday"
day 7 = "Sunday"
day x = "The week has only 7 days!"
```

You'll get the following:

```
*Main> day 7
"Sunday"
*Main> day 0
"The week has only 7 days!"
*Main> day 3
"Wednesday"
```

When day is called, the matching begins from the bottom. When there is a match, the corresponding body is chosen. Note that the function contains a default pattern. If you do not put a default pattern and if the function's parameter does not fall into any defined pattern, then you will get an exception. If you comment out or delete the last line of the function, you get the following result:

```
*Main> day 0
"*** Exception: Days.hs:(2,1)-(8,16): Non-exhaustive patterns
in function day
```

This function can be written using an if..then..else statement, but it would be pretty difficult to follow. This statement is self-explanatory: *if* the condition is met, *then* choose the value from the first branch, *else* choose the value from the second branch. Here's a short example:

```
numbers :: (Integral a) => a -> String
numbers x =
 if x < 0 then "negative"
 else "positive"

*Main> numbers 8
"positive"
*Main> numbers  (-1)
"negative"
```

For negative numbers, don't forget to put them inside parentheses.

Case Expressions

Case expressions are simple. The general syntax is as follows:

```
case expression of pattern -> result
                   pattern -> result
                      ...
```

Let's write the day function using case, as shown here:

```
day :: (Integral a) => a -> String
day x = case x of 1 -> "Monday"
                  2 -> "Tuesday"
                  3 -> "Wednesday"
```

```
        4 -> "Thursday"
        5 -> "Friday"
        6 -> "Saturday"
        7 -> "Sunday"
        _ -> "The week has only 7 days!"
```

Then write the following:

```
Prelude> :l Day.hs
[1 of 1] Compiling Main            ( Day.hs, interpreted )
Ok, one module loaded.
*Main> day 7
"Sunday"
*Main> day 10
"The week has only 7 days!"
```

Guards

You can use guard to test whether a value has a certain property. Guards are alternative to else..if statements, making the code easier to write and follow. Let's continue with a sign example, shown here:

```
sign :: (RealFloat a) => a -> String
sign x
  | x < 0     = "negative"
  | x == 0    = "zero"
  | otherwise = "positive"
*Main> sign 3.5
"positive"
*Main> sign (-7)
"negative"
```

As you can see in this example, to use guards, you mark them with *pipes*. The evaluation begins with the bottom expression and continues until a match is found. Note that we have the default case marked by the otherwise keyword; any value that does not meet any of the previous conditions will get the default. You can think of guards as Boolean expressions, where otherwise is always evaluated as True. Pay attention to what's after the parameters in the function definition. Note that you don't put an equal sign.

Clauses

In this section, you'll learn about the where clause and let..in clause.

Let's begin with the where clause. As example, think about the quadratic equation defined as $ax^2 + bx + c = 0$. The solutions of the equation depend on a *discriminant* computed as $\Delta = b^2 - 4ac$. If $\Delta > 0$, you will obtain two real solutions: x_1, x_2. If $\Delta = 0$, you will obtain two real identical solutions: $x_1 = x_2$. Otherwise, the equation does not have real solutions. You can think of the three parameters a,b,c as a triple and the solutions as a pair. Observe that you need Δ in more parts of the algorithm. You can resolve it as follows:

```
quadraticEq :: (Float, Float, Float) -> (Float, Float)
quadraticEq (a, b, c) = (x1, x2)
  where
    x1 = (-b - sqrt delta) / (2 * a)
    x2 = (-b + sqrt delta) / (2 * a)
    delta = b * b - 4 * a * c
```

Let's test this.

```
*Main> quadraticEq (1, 2, 1)
(-1.0,-1.0)
*Main> quadraticEq (1, 1, 1)
(NaN,NaN)
```

```
*Main> quadraticEq (1, 4, 0)
 (-4.0,0.0)
```

The names defined in the where clause are used only in the function, so they will not affect other functions or modules. Be careful about the indentation; all the names should be aligned properly. Do not use the Tab key to add large spaces. Also, the names defined in the body of a function pattern will not be visible by the body of the same function for another pattern. If you need a name to be visible in all patterns, you need to declare it globally. You can also define a function in the where clause.

Another useful clause is let..in. Suppose you want to compute the volume of a quadrilateral pyramid. You know that $V = \dfrac{A_b \cdot h}{3}$, where A_b is the area of the basis. You can proceed as follows:

```
pyramivVol :: (RealFloat a) => a -> a -> a
pyramivVol l h =
  let area = l^2
  in  (area * h)/3
```

The names defined in the let clause are available just in the in. You can find the volume also using the where clause. The difference between them is that the let bindings are expressions themselves, while where bindings are syntactic constructs. In previous chapters, you used the let.. in clause to define functions and constants in GHCi. There, you ignored the in part, which means that the names were available through the entire session. This clause can be used almost anywhere, not only in functions.

```
Prelude> 5 * (let a = 2 in a^2) + 7
27
Prelude> [let cube x = x^3 in (cube 6, cube 3)]
[(216,27)]
Prelude> (let x = 100; y = 200 in x+y, let l="Anne ";
f = "Scott" in l ++ f)
(300,"Anne Scott")
```

The in part can miss when the visibility of the names are predefined. This clause can be also used in list comprehension, inside a predicate, but the names will be visible only in that predicate.

Lambda Expressions

There are times when you need to use a function just once in your entire application. To not complete with names, you can use anonymous blocks called *lambda expressions*. A function without definition is called a *lambda function*, and it is marked by the \ character. Let's take a look:

```
main = do
    putStrLn "The square of 2 is:"
    print ((\x -> x^2) 2)
```

Inside print, we defined the expression x->x^2 and call it for the value 2. The output is as follows:

```
The square of 2 is:
4
```

Infix Functions

In Haskell, the functions are called by typing the name of the function, followed by the arguments. But there are functions that don't follow this rule, such as mathematical operators. Actually, you can call an operator, followed by the two arguments, but this is unnatural. Therefore, a function that stands between its two arguments is called an *infix function*.

```
Prelude> (+) 2 2
 4
Prelude> (*) 5 6
 30
```

Let's define an infix function, as shown here:

```
Prelude> let concatAndPrint a b = putStrLn $ (++) a b
 Prelude> concatAndPrint "abc" "def"
 abcdef
 Prelude> "abc" `concatAndPrint` "def"
 abcdef
```

Note that the infix function is marked by `` signs and is between its two arguments. Usually, an infix function is used with two parameters.

Higher-Order Functions

"A higher-order function is a function that takes other functions as arguments or returns a function as result."[1] Let's define the following function:

```
multiplyList m [] = []
multiplyList m (y:ys) = m*y : multiplyList m ys

*Main> multiplyList 3 [2, 5, 7]
[6,15,21]
```

Note that it takes two inputs, a number and a list, and the output is a list resulting from multiplying the number with the elements.

A higher-order function is multiplyListBy3.

```
multiplyListBy3 = multiplyList 3

*Main> multiplyListBy3 [10, 20, 30]
[30,60,90]
```

---

[1]https://wiki.haskell.org/Higher_order_function

The function `multiplyListBy3` takes now one input, namely, a list, because you know that `m=3` when `multiplyList` is called inside `multiplyListBy3`.

Summary

In this chapter, you learned how to write source code files and how to use them in GHCi. In addition, you learned how to use pattern matching and guards in functions and what the difference is between them. You also saw that in Haskell you can use a case expression. Next, you learned how to use clauses in functions. You also learned about lambda expressions and how you can write a function that is used just once in your entire application. Finally, you worked with infix functions and higher-order functions.

References

1. Haskell/control structures, `https://en.wikibooks.org/wiki/Haskell/Control_structures`

2. Guard (computer science), `https://en.wikipedia.org/wiki/Guard_(computer_science)`

3. P. Hudak, J. Peterson, and J. Fasel, *A Gentle Introduction to Haskell 98* (1999)

4. R. Bird, *Introduction to Functional Programming Using Haskell*, vol. 2 (Prentice Hall Europe, 1998)

CHAPTER 8

Recursion

In the previous chapter, you learned about functions in Haskell. Many times, in real-world applications you'll work with functions that recall themselves, which is a process called *recursion*. In this chapter, you will learn what recursion is, see some concrete examples, and learn how you implement them in Haskell and how to apply recursion on lists.

> *Recursion in computer science is a method of solving a problem where the solution depends on solutions to smaller instances of the same problem (as opposed to iteration).*
>
> —Wikipedia[1]

Let's start with a simple example. Do you remember the factorial function from math class? Well, it is defined as follows:

$$f : \mathbb{N} \to \mathbb{N}, f(n) = \begin{cases} 1 & , n = 0 \\ n \cdot f(n-1), & n > 0 \end{cases}$$

This function is also known as `n!`. The first branch represents an *edge condition*, which tells when the function terminates. It is important because it will finish the recursive call; otherwise, the execution will enter into an infinite loop.

---

[1]Recursion (computer science), `https://en.wikipedia.org/wiki/Recursion_(computer_science)`

© Stefania Loredana Nita and Marius Mihailescu 2019
S. L. Nita and M. Mihailescu, *Haskell Quick Syntax Reference*,
https://doi.org/10.1007/978-1-4842-4507-1_8

53

Returning to the factorial function, you write it in Haskell as follows:

```
fact 0 = 1
fact n = n * fact (n - 1)
```

After writing the previous two lines into a file called Fact.hs and loading the file, let's test it.

```
*Main> fact 5
120
*Main> fact 0
1
*Main> :t fact 5
fact 5 :: (Eq p, Num p) => p
```

When Haskell needs to decide which function definition to choose, it starts with the foremost definition and picks up the first one that matches. Therefore, the order of definitions in a recursive function is important. If you switched the definitions, you would never get a result, because Haskell would always be stuck in the first definition, fact n = n*fact(n-1). This fact leads us to the following note.

Note   Always start a recursive function body with the definitions for the edge conditions. In other words, begin with the particular cases before the general case.

If you call fact for a negative number, an infinite loop will be encountered. Why is that? This happens because Haskell will always choose the general definition, because it will never reach the edge condition. Anticipating a little, you can use the function error from Prelude to report an error (in this example, the function fact cannot be applied on negative integers or any other value that's not positive). You can improve the function fact as follows:

```
fact 0 = 1
fact n | n > 0 = n * fact (n - 1)
       | otherwise = error "Wrong input"
```

Now, for example, if you call fact for -5 or 3.4, you get the following:

```
*Main> fact (-5)
*** Exception: Wrong input
CallStack (from HasCallStack):
  error, called at Fact.hs:3:22 in main:Main
*Main> fact 3.5
*** Exception: Wrong input
CallStack (from HasCallStack):
  error, called at Fact.hs:3:22 in main:MainRecursive Functions
  on Lists
```

Handling for and while Loops from Imperative Languages

In Haskell, you don't have for or while loops like in imperative languages; instead, you declare what something *is* instead of saying *how* to get it. Therefore, recursion is even more important because you use it to say what something is.

In functional languages, these loops are expressed as computations over lists. Usually, the following three functions can be used to replace for/while loops:

- map applies a function given as a parameter to every element of a list.

- foldl goes through the entire list from left to right.

- foldr goes through the entire list from right to left.

Of course, these can be combined with other functions (predefined or your own functions) and used on lists to create algorithms that are more complex.

Further, let's say you have a list of positive integers and you want to compute every element's factorial. You already saw a fact function defined in the previous section, which computes the factorial for one number. To apply it on the numbers of a list, you proceed as follows:

```
*Main> map fact [2, 9, 8]
[2,362880,40320]
```

How simple is that? In this way, you eliminated the need for the while/for loop.

You will learn more about foldl and foldr in Chapter 15.

Recursion on Lists

Let's begin this section with a simple example: list creation itself. You already know that a list can be empty and you can add an element into a list using the : operator. For example, the list [1,2,3,4] is created as 1:(2:(3:(4:[]))). Here, you used : recursively (remember that an operator is actually a function).

The next function you'll look at is reverse. You can define your recursive recursiveReverse as follows:

```
recursiveReverse :: [a] -> [a]
recursiveReverse [] = []
recursiveReverse (x:xs) = recursiveReverse xs ++ [x]
```

Let's see how it works:

```
*Main> recursiveReverse [3,2,6,7]
[7,6,2,3]
```

Remember that reverse reverses a list. In the function's body, a reversed empty list is the empty list itself. On the last line, you concatenate the first element of the list given as a parameter to the end of an existing list. For this example, it works like this:

```
recursiveReverse [3,2,6,7]
recursiveReverse [2,6,7] ++ [3]
(recursiveReverse [6,7] ++ [2]) ++ [3]
((recursiveReverse [7] ++ [6]) ++ [2]) ++ [3]
(((recursiveReverse [] ++ [7]) ++ [6]) ++ [2]) ++ [3]
```

On the last line, the call reaches the edge condition. Further, it continues with the following:

```
((([] ++ [7]) ++ [6]) ++ [2]) ++ [3]
(([7] ++ [6]) ++ [2]) ++ [3]
([7,6] ++ [2]) ++ [3]
[7,6,2] ++ [3]
[7,6,2,3]
```

That's it. Next, you implement the recursive version of the filter function.

```
recursiveFilter :: (a -> Bool) -> [a] -> [a]
recursiveFilter condition []     = []
recursiveFilter condition (x:xs)
        | condition x          = x : recursiveFilter condition xs
        | otherwise       = recursiveFilter condition xs
```

You call it as follows:

```
*Main> recursiveFilter (<10) [2, -9, 4, 20, 0, 100]
[2,-9,4,0]
*Main> import Data.Char
*Main Data.Char> recursiveFilter (isLetter) "Ann has 5 apples."
"Annhasapples"
```

To use the isLetter function, you need to import the Data.Char module.

Other functions on lists that can be defined recursively are length, repeat, maximum, minimum, take, and `elem`.

```
recursiveLength :: [a] -> Int
recursiveLength [] = 0
recursiveLength (x:xs) = 1 + recursiveLength xs
```

```
recursiveReverse :: [a] -> [a]
recursiveReverse [] = []
recursiveReverse (x:xs) = recursiveReverse xs ++ [x]
```

```
recursiveRepeat :: a -> [a]
recursiveRepeat x = x:recursiveRepeat x
```

```
recursiveMaximum :: (Ord a) => [a] -> a
recursiveMaximum [] = error "Empty list"
recursiveMaximum [x] = x
recursiveMaximum (x:xs) = max x (recursiveMaximum xs)
```

```
recursiveMinimum :: [Int] -> Int
recursiveMinimum (x:[]) = x
recursiveMinimum (x:xs) = x `min` recursiveMinimum xs
```

```
recursiveTake :: (Num i, Ord i) => i -> [a] -> [a]
recursiveTake n _
    | n <= 0   = []
recursiveTake _ []     = []
recursiveTake n (x:xs) = x : recursiveTake (n-1) xs
```

```
recursiveElem :: (Eq a) => a -> [a] -> Bool
recursiveElem a [] = False
recursiveElem a (x:xs)
    | a == x    = True
    | otherwise = a `elem` xs
```

Feel free to use and test these functions.

Pattern Matching and Recursion

In this section, you will learn about a pattern matching technique that allows you to write a standard skeleton for recursive functions on lists.

Let's start with a general definition.

```
recursiveFunction [] = -- this is the edge condition
recursiveFunction (x:xs) = -- this is the general case
```

The edge condition is not allowed to include recursive calls; otherwise, it would lead to an infinite loop. The edge condition *must* return a value, as you saw in the previous sections. The type of the returned value needs to be the same as the type annotation in the function. Here's an example:

```
recursiveReverse :: [a] -> [a]
```

This definition means the edge condition should have the type [a], so the "most particular" case for [a] is the empty list [].

```
recursiveReverse [] = []
```

For the recursion, the recursive function needs to call itself like this:

```
recursiveFunction (x:xs) = function x recursiveFunction xs
```

or like this:

```
recursiveFunction (x:xs) = recursiveFunction xs function x
```

function assures the chaining of recursive calls. The order of recursiveFunction and function depends on the effective specifications. To make it more intuitive, here is how to use function as an operator:

```
recursiveFunction (x:xs) = x `function` recursiveFunction xs
```

Don't forget to check the annotation type of the recursive function because `function` needs to act accordingly.

For example, in recursiveReverse, `function` is actually the operator ++.

```
recursiveReverse (x:xs) = recursiveReverse xs ++ [x]
```

Finally, you can obtain the recursive version of recursiveReverse, as defined in the previous section.

Summary

In this chapter, you learned about the following:

- What recursion is and why it is important

- How to handle for and while loops from imperative languages by using recursion in the functional language Haskell

- How to write some common functions recursively on lists

- How to write a general skeleton of a recursive function using pattern matching

References

1. R. Hinze and J. Jeuring, "Generic Haskell: Practice and Theory." *Generic Programming* (Springer, 2003), 1-56

2. C. McBride, "Faking It Simulating Dependent Types in Haskell," *Journal of Functional Programming*, 12(4–5), 375–392 (2002)

3. Recursion, `https://en.wikipedia.org/wiki/`
    `Recursion`

4. Haskell/recursion, `https://en.wikibooks.org/`
    `wiki/Haskell/Recursion`

5. Recursion, `http://learn.hfm.io/recursion.html`

6. Recursive functions on lists, `https://www.`
    `futurelearn.com/courses/functional-`
    `programming-haskell/0/steps/27211`

CHAPTER 9

List Comprehension

Chapter 6 introduced lists, and now you will learn another way to represent a list. Do you remember how sets are represented using mathematical symbols? Well, you will do that with lists. Further, you will learn more complex functions that you can apply on lists.

Introduction

Let's say you have the set $A = \{x \in N | 5 \leq x \leq 10\}$. If you pay a little attention, you will observe that in natural terms A is a list. Haskell provides you with a way to represent such lists, as shown here:

```
Prelude> let set = [x | x <- [5..10]]
Prelude> set
[5,6,7,8,9,10]
```

This type of representation is called *list comprehension*, and x is called a *generator*.

This can be complicated. Let's say you want the numbers between 100 and 200 that are divisible by 17. Note that you can add a condition (or a predicate) called a *guard*.

```
Prelude> [x | x <- [100..200], x `mod` 17 == 0]
[102,119,136,153,170,187]
```

© Stefania Loredana Nita and Marius Mihailescu 2019
S. L. Nita and M. Mihailescu, *Haskell Quick Syntax Reference*,
https://doi.org/10.1007/978-1-4842-4507-1_9

What if you want those numbers divisible with 17 *and* 10? You do that like this:

```
Prelude> [x | x <- [100..200], x `mod` 17 == 0, x`mod` 10 == 0]
[170]
```

Observe that we added two predicates, separated by a comma.

But you also can do it like this:

```
Prelude> [x | x <- [100..200], x `mod` 17 == 0 && x`mod` 10 == 0]
[170]
```

You can add more variables like this:

```
Prelude> [ x+y | x <- [1,7,12], y <- [5,9,14]]
[6,10,15,12,16,21,17,21,26]
Prelude> [ x+y | x <- [1,7,12], y <- [5,9,14,17]]
[6,10,15,18,12,16,21,24,17,21,26,29]
```

In this case, every element from the first list is summed up with every element from the second list. It is similar to a Cartesian product, but in this case you sum up the elements of pairs in the Cartesian product.

Let's add another restriction to this example, as shown here:

```
Prelude> [ x+y | x <- [1,7,12], y <- [5,9,14,17], x+y >20]
[21,24,21,26,29]
```

These operations work the same on the strings.

```
Prelude> let l1 = ["my", "your"]
Prelude> let l2 = ["book", "pencil", "PC"]
Prelude> [elem1 ++ " " ++ elem2 | elem1 <- l1, elem2 <- l2]
["my book","my pencil","my PC","your book","your pencil","your PC"]
```

An important aspect about lists is the _ symbol. Here's an example:

```
Prelude> let numbers = [1,3..100]
Prelude> [0 | _ <- numbers]
[0,0,0,0,0,0,0,0,0,0,0,0,0,0,0,0,0,0,0,0,0,0,0,0,0,0,0,0,0,0,0,
0,0,0,0,0,0,0,0,0,0,0,0,0,0,0,0,0,0,0]
```

This means every element of the list is replaced by 0. Instead of _,
you could use a variable, but it is more convenient in this way, as
syntactic sugar. For short, *syntactic sugar* is just another representation for
expressions written in an analytic way and does not add functionality. The
main purpose of syntactic sugar is to make the code more readable.

You can use an if..then..else statement in a list comprehension.
Let's say you want to check whether the positive numbers in a list are odd
or even.

```
Prelude> parity list = [ if x `mod` 2 == 0 then "even" else
"odd" | x <- list, x >= 0]
Prelude> parity [-100,-97..25]
["even","odd","even","odd","even","odd","even","odd"]
```

This is a little confusing because you don't know which are the
numbers. Let's improve it.

```
Prelude> parity list = [ if x `mod` 2 == 0 then show x ++ "
even" else show x ++ " odd" | x <- list, x >= 0]
Prelude> parity [-100,-97..25]
["2 even","5 odd","8 even","11 odd","14 even","17 odd","20
even","23 odd"]
```

Now it is better. The show function converts its parameter into a
String.

Other Functions on Lists

Now that you know how to work with functions, you'll write your own functions to apply on lists. You will start by using some other functions from the Data.List module (all other functions on lists that you have learned about so far belong to this module), and then you will sort elements on a list of integers.

In the first example, you will use the predefined map function. The general definition of map is as follows:

```
map :: (a -> b) -> [a] -> [b]
```

The statement map f list will produce a new list obtained by applying f to every element of list. Write a function double in the Double.hs file with the following statements:

```
double :: Integer -> Integer
double x = 2*x
```

Then, load the file and apply the double function to the elements of a list.

```
Prelude> :load Double
[1 of 1] Compiling Main              ( Double.hs, interpreted )
Ok, one module loaded.
*Main> map double [-8,-3..30]
[-16,-6,4,14,24,34,44,54]
```

Next, you want to check whether an element of a list is even and whether it is greater than 20. You can use the check function.

```
check :: Integer -> Bool
check x =
  x `mod` 2 == 0 && x > 20
```

Further, you can use the any function that tells you whether there are elements in the list that accomplish a condition.

```
*Main> any check [1,2,3,4]
False
*Main> any check [1,2,3,4, 22]
True
```

A similar function is all, which tells you whether all the elements of the list accomplish a condition.

```
*Main> all check [1,2,3,4]
False
*Main> all check [1,2,3,4, 22]
False
*Main> all check [22, 24]
True
```

The following are other useful functions for lists:

- find returns the first element of the list that satisfies a predicate, or Nothing otherwise.

- filter returns all elements of the list that satisfy a condition.

- elemIndices/elemIndex returns all indices of the elements equal to a given item or the first index of the element equal to a given item or Nothing if there is no such element.

- findIndex/findIndices returns all indices of the elements that make up a predicate or the first index of the element that makes up a predicate or Nothing if there is no such element.

- The zip family of functions creates tuples from lists based on different criteria.

Go ahead and practice these functions.

Next, let's sort a list. For this example, you can use the *quicksort* technique. Quicksort takes an element as a pivot and reorders the list so that all elements with a lesser value than the pivot move onto the left side of the pivot and the elements with a greater value than the pivot move to the right side of the pivot (assuming you are sorting a list in increasing order). This can be done recursively easily.

```
qsort :: Ord a => [a] -> [a]
qsort [] = []
qsort (y:ys) = (qsort ls) ++ [y] ++ (qsort gt)
    where
        ls = filter (< y) ys
        gt = filter (>= y) ys
```

This works as follows:

```
*Main> qsort [10, 1, 5]
[1,5,10]
*Main> qsort [10, 1, 5, -7, 0]
[-7,0,1,5,10]
```

Summary

In this chapter, you learned about the following:

- How to represent a list in a comprehension form

- How to apply your own functions on lists

- Other useful functions applied on lists

- How to sort a list, using the quicksort algorithm

References

1. T. H. Cormen, C. E. Leiserson, R. L. Rivest, and C. Stein, *Introduction to Algorithms* (MIT Press, 2009)

2. A. Nunes-Harwitt, M. Gambogi, and T. Whitaker, "Quick-Sort: A Pet Peeve" in proceedings of the 49th ACM Technical Symposium on Computer Science Education, pp. 547–549 (ACM, 2018)

3. M. Lipovaca, *Learn You a Haskell for Great Good! A Beginner's Guide* (No Starch Press, 2011)

4. M. Aslam, *Functional Programming Language-Haskell* (2003)

5. Data.List, http://hackage.haskell.org/package/base-4.12.0.0/docs/Data-List.html

6. List comprehension, https://wiki.haskell.org/List_comprehension

CHAPTER 10

Classes

In the previous chapters, you created your own types. Now, it's time to learn what classes are in Haskell and how to work with them. In this chapter, you'll learn about the standard classes and then how to create your own classes.

Standard Classes

In Haskell, the most common standard classes are the following ones:

- Eq is used when you work with == (is equal) and /= (is not equal).

- Ord is used when you work with the operators <, <=, >, and >= and the functions min, max, and compare.

- Enum is used in enumerations and lets you use syntax such as [Red .. Yellow].

- Read includes the function read, which takes a string as a parameter and parses it into a value.

- Show includes the function show, which takes a value as a parameter and converts it to a string.

- Bounded is used in enumerations and includes the functions minBound and maxBound.

© Stefania Loredana Nita and Marius Mihailescu 2019
S. L. Nita and M. Mihailescu, *Haskell Quick Syntax Reference*,
https://doi.org/10.1007/978-1-4842-4507-1_10

Do you remember in Chapter 4 when you used the command deriving followed by one of the previous classes? You did that because you wanted your type to use functions that actually belong to these classes. For example, to compare two dates of type DateInfo, you used deriving Eq. That allowed you to write myDate2 == myDate1.

The Eq Class

The following is the minimal definition of Eq from Prelude[1]:

```
class  Eq a  where
  (==), (/=) :: a -> a -> Bool
  x /= y     = not (x == y)
  x == y     = not (x /= y)
```

This says that if type a is an instance of Eq, it must support == and /=. The operators == and /= are called *class methods*, and they are defined in *terms of each other*. This means that a type in Eq should provide the definition for one of them, with the other being deduced automatically. Let's take a look at ==, shown here:

```
(==) :: (Eq a) => a -> a -> Bool
```

This says that the == has the type a->a->Bool for every type a that is an instance of the class Eq.

---

[1]http://hackage.haskell.org/package/base-4.12.0.0/docs/Prelude.
html#t:Eq

When you already have a type, you can make it an instance of a class, which is a process called *instance declaration*.

```
data MyType = MyType {stringVal :: String, intVal :: Integer}

instance Eq MyType where
    (MyType s1 i1) == (MyType s2 i2) = (s1 == s2) && (i1 == i2)

*Main> MyType "apples" 2 == MyType "apples" 2
True
*Main> MyType "apples" 2 /= MyType "apples" 2
False
```

In the previous code, we defined the type MyType and then made it an instance of Eq. Further, we said that two values of type MyType are equal (==) if the string of the first value is equal to the string of the second value *and* the integer of the first value is equal to the integer of the second value. It worked. Observe that you use /= even if you don't define it in the instance declaration of MyType; this is because Haskell knows how to automatically deduce it from the definition of ==.

Inheritance

All types you define should be an instance of Eq and even Show or Ord. When the definitions of methods of these classes are evident, then you can use deriving, as in Chapter 4. In this way, you avoid having to write complex definitions. But you can derive just some of the standard classes: Eq, Show, Ord, Enum, Bounded, Read.

A class can inherit another class. For example, a brief definition of Ord from Prelude[2] is as follows:

```
class  (Eq a) => Ord a  where
   compare                :: a -> a -> Ordering
   (<), (<=), (>=), (>) :: a -> a -> Bool
   max, min               :: a -> a -> a
```

The => sign says that Ord inherits Eq, which means that a type that is an instance of Ord is also an instance of Eq. Thus, it must also implement == and /=. A class can inherit multiple classes.

```
class  (Num a, Ord a) => Real a  where
   -- | the rational equivalent of its real argument with full
   precision
   toRational          ::  a -> Rational
```

This definition is from Prelude,[3] and the multiple inheritance is marked by the inherited classes in the parentheses.

Figure 10-1 presents the hierarchy of classes in Haskell, taken from the Haskell report[4]. The names of classes are in bold, and the instances are regular font. In every ellipse, -> means function, and [ ] means list. The arrows between ellipses show the inheritance relationships, indicating the inheriting class.

---

[2]http://hackage.haskell.org/package/base-4.12.0.0/docs/Prelude.html#t:Ord
[3]http://hackage.haskell.org/package/base-4.12.0.0/docs/Prelude.html#t:Real
[4]https://www.haskell.org/onlinereport/basic.html#standard-classes

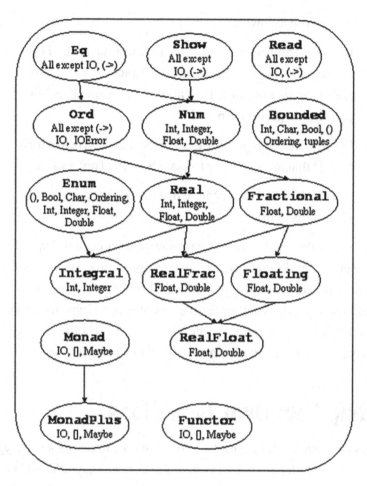

Figure 10-1. *Hierarchy of classes in Haskell [4]*

You can add *constraints* to your types, in the following keywords :

- instance: You declare parametrized types.

- class: You can add constraints, different from the ones in the class definition, in method signatures.

- data: This constrains the constructor signatures.

Further, note the following:

- In our examples, we used three definitions marked by the keywords class, data, and instance. In fact, these are separate, and there isn't any rule that specifies how you should group them.

- Classes are not types; they are categories of types, which means that an instance of a class is a type (not a value!).

- You can't define an instance of a class from type synonyms defined with the keyword type.

In the examples in this section, you saw that you can test whether two values are equal for different types. For example, you can test whether two integer values are equal, you can test whether two string values are equal, and you can test whether two values of type MyType are equal. So, you can apply the == operator in different types. This behavior is known as *overloading* (or less commonly known as *ad hoc polymorphism*).

Creating Your Own Type Class

Now, let's create a simple example. You'll see how to define a data type called Animal, which can be Cat, Dog, or Parrot. Then, you'll tell Haskell which is equal to which and how to show them.

```
data Animal = Cat | Dog | Parrot

instance Eq Animal where
  Cat == Cat = True
  Dog == Dog = True
  Parrot == Parrot = True
  _ == _ = False
```

```
instance Show Animal where
  show Cat = "In ancient times cats were worshipped as gods;
  they have not forgotten this."
  show Dog = "It is Human's best friend."
  show Parrot = "It repeats everything you say."
```

Now let's test it.

```
*Main> Cat == Cat
True
*Main> Cat == Parrot
False
*Main> show Cat
"In ancient times cats were worshipped as gods; they have not
forgotten this."
```

Advanced Type Classes

Now, let's suppose you want to define a class Set, which allows you to add an element to the set and test whether a value is an element of that set. You might proceed as follows:

```
class Set a where
    add :: a -> b -> a
    isElem :: a -> b -> Bool

instance Set [c] where -- List becomes instance of Set
    add xs x = x:xs
    isElem = flip elem
```

But this definition is not correct, because type b in the method definitions is unknown, so Haskell will not know from where to take it. A way to correct this is to use type classes with multiple parameters, which allows you to integrate b into the type of the class.

```
{-# LANGUAGE FlexibleInstances #-}
{-# LANGUAGE MultiParamTypeClasses #-}
class Eq b => Set a b where
  add :: a -> b -> a
  isElem :: a -> b -> Bool

instance Eq c => Set [c] c where
  add = flip (:)
  isElem = flip elem
```

The previous code uses *language extensions*. This is marked by the LANGUAGE pragma.

```
{-# LANGUAGE <Extension>, <Extension> #-}
```

Language extensions enable features of Haskell that are useful in certain contexts. In this example, we needed multiparameter type classes. In addition, we used flexible instances, which allows a type parameter to occur twice in a type class instance (for Set [c] c).

Still, the definition is not entirely correct because it will lead to ambiguities. Your intuition might tell you that the type of set will provide the type of its elements, so the type of elements *depends on* the type of the set. For example, if a is [p], then b is p. To make things clear for the compiler, you add another language extension, called *functional dependency*:

```
{-# LANGUAGE FunctionalDependencies #-}
 class Eq b => Set a b | a -> b where ...
```

In the previous definition, | a -> b means "a uniquely identifies b." In other words, for a given b, it will be just one a.

You can add as many constraints as you want (of course, they need to make sense) to a class definition, and in a multiparameter class you can use more than two parameters.

Maybe, Just, and Nothing

In Prelude, Maybe is a type that has two constructors: Just a or Nothing. In situations in which a type has more constructors, it must be constructed with only one constructor, so Maybe is constructed with Just a (a can be any type) or Nothing. Let's take a closer look.

- Nothing: When constructed with Nothing, Maybe is defined as a constant that becomes a member of Maybe a, for all types a, because Nothing doesn't take a parameter type;.

- Just a: When constructed with Just, Maybe is used as a type parameter a; in this scenario, Just behaves as a function from a to Maybe a, meaning its type is a-> Maybe a.

When Maybe is used with pattern matching, two patterns are necessary, one for each constructor. Here's an example:

```
case maybeExample of
    Nothing -> "This is the Noting constructor."
    Just a -> "This is the Just constructor, with value "
    ++ (show a)
```

Maybe is mostly used to extend types with the Nothing value, i.e., the absence of a value. This approach prevents errors. In other programming languages, the "no value" is treated with a NULL reference.

Functor

The type class Functor provides a way to make the operations from a base type work with a new type constructed by transforming the base type into the new one. Functor contains the function fmap, which is used to map the function that takes values from the base type with functions that take values from the new type.

Functor can be combined with Maybe.

```
case maybeExample of
  Nothing  -> Nothing
  Just a -> Just (f a)
```

In the first branch, there is no value, so return Nothing; in the second branch, there is the value a, so apply the function f to a.

For example, if you work with a value valueI of type Maybe Integer and a function f that goes from an integer to other integer (in other words, Int -> Int), then you can use fmap f valueI to apply f directly on the Maybe Integer value. When using Maybe, it's safe to think there is nothing to worry about if it didn't get a value.

Summary

In this chapter, you learned the following:

- What standard classes Haskell has and what the class hierarchy looks like

- How to inherit standard classes

- How to define your own type classes and how to avoid common mistakes in class definitions

- What language extensions are and when you can use them

- What Maybe, Just, and Nothing are and how can you use them with Functor

References

1. A. Serrano Mena. *Beginning Haskell: A Project-Based Approach* (Apress, 2014)

2. B. Heeren and J. Hage, "Type class directives," International Workshop on Practical Aspects of Declarative Languages (Springer, 2005)

3. W. Kahl and J. Scheffczyk. "Named Instances for Haskell Type Classes," in proceedings of the 2001 Haskell Workshop, UU-CS-2001-23 (Tech. Rep., 2001)

4. Predefined types and classes, https://www.haskell.org/onlinereport/basic.html#standard-classes

5. Prelude, http://hackage.haskell.org/package/base-4.12.0.0/docs/Prelude.html

6. Type classes and overloading, https://www.haskell.org/tutorial/classes.html

7. Standard Haskell classes, https://www.haskell.org/tutorial/stdclasses.html

8. Type classes, https://www.schoolofhaskell.com/school/starting-with-haskell/introduction-to-haskell/5-type-classes

9. Language extensions, https://wiki.haskell.org/Language_extensions

CHAPTER 11

Pattern Matching

In Chapter 7, you learned the basics of pattern matching, which is used with functions. In this chapter, you'll learn more about the details of pattern matching.

Basically, *pattern matching* means matching values with patterns and binding variables with the matches that succeeded.

Let's take a closer look at the map function. The type signature and the definition are as follows:

```
map _ []     = []
map f (x:xs) = f x : map f xs
```

Here, you can identify four types of patterns.

- f: This pattern matches basically anything and binds the f to anything that fits in.

- (x:xs): This pattern matches a list with at least one element; the list contains something (which is bounded to x) created (with (:) operator) from something else (bounded to xs).

- []: This pattern matches an empty list and does not bind anything.

- _: This pattern matches everything but does not bind anything.

© Stefania Loredana Nita and Marius Mihailescu 2019
S. L. Nita and M. Mihailescu, *Haskell Quick Syntax Reference*,
https://doi.org/10.1007/978-1-4842-4507-1_11

In the expression (x:xs), the x and xs can be considered subpatterns that match parts of a list, matching with everything that respects the types in the type signature of map. In other words, x matches with anything of type a, and xs matches with anything of type [a], in particular with an empty list. Therefore, a list with one element matches (x:xs).

Pattern matching is useful in the following situations:

- **Recognizing values**: For example, map's definition says that when an empty list is the second parameter, the result will be an empty list. In other words, it is chosen as the first branch.

- **Binding variables to identified values**: For example, f, x, and xs from the previous definition are mapped to the arguments of map when called, and the second branch is chosen. Binding can be seen as a side effect of the fact that variables names are used as patterns, as using _ and [] suggests.

- **Sectioning values into more parts**: For example, the (x:xs) expression binds x with the head of a list and xs with the tail of a list.

Further, let's consider a function that duplicates the first element in a list. It should look like this:

```
g (y:ys) = y:y:ys
```

Observe that y:ys appears both on the left side and on the right side. To make the code easy to follow, you can use the *as-pattern* (@), which allows you to write y:ys just once.

```
g s@(y:ys) = y:s
```

Even the subpattern can fail, i.e., y:ys. The as-pattern always matches. Another advantage is that it can be faster because in the first version, the y:ys is reconstructed, instead of reusing the value to which it matched.

The result of a pattern matching process can have one of the following states:

- **Success**: When a pattern matching process succeeds, the variables are bounded with the arguments.

- **Fail**: When in a pattern matching process an equation fails, then the matching process moves to the next equation, and so on. If all equations fail, then a runtime error occurs.

- **Divergence**: When a pattern matching process diverges, it means that a value needed in the pattern contains an error.

Pattern Matching and Constructors

Not all functions are allowed to be used in pattern matching. In pattern matching you use just constructors—those functions that construct data types.

Let's examine the following piece of code:

```
data MyData = Zero | Double Int

g :: MyData -> Int
g Zero = 0
g (Double x) = 2*x
```

Here, Zero and Double are constructors for the MyData type. You can use them to *pattern match* Zero with the 0 value of the Int type and *bind* a value constructed with Double from MyData with its Int double.

Pattern matching works on lists, because on a naïve implementation you can interpret lists as defined with the data keyword.

```
data [a] = [] | a : [a]
```

Here, the type constructors for the list are the empty list and the (:) operator.

Note that the previous definition is not actually correct; we used it as an intuitive explanation of the reason why the pattern matching works on lists. In fact, in Haskell, lists are important, and they have a much more complex construction.

Tuples have a similar explanation.

Pattern matching is useful in records. For a comprehensive explanation of records, refer to Chapter 4.

Uses of Pattern Matching

You can use pattern matching in the following scenarios:

- Equations (see the map example)
- let expressions

  ```
  Prelude> y = let (x:_) = map (+8) [5,6] in x * 3
  Prelude> y
  39
  ```

- where clauses

  ```
  Prelude> y = x * 3 where (x:_) = map (+8) [5,6]
  Prelude> y
  39
  ```

- Lambda abstractions

```
Prelude> switch = \(a,b) -> (b,a)
Prelude> switch (5,6)
(6,5)
```

- List comprehension

```
associate :: Eq a => a -> [(a, t)] -> [t]
associate character xs = [y | (x,y) <- xs ,
x == character]
```

```
Prelude> associate 'D' [('D', 15), ('B', 0), ('D', 35),
('D', 100)]
[15,35,100]
```

- do blocks

```
firstLetter = do
  (x:_) <- getLine
  putStrLn [x]
```

Summary

In this chapter, you learned the following:

- What pattern matching is

- How pattern matching and constructors are related

- When to use pattern matching

References

1. M. Lipovaca, *Learn You a Haskell for Great Good! A Beginner's Guide* (No Starch Press, 2011)

2. B. O'Sullivan, J. Goerzen, and D. B. Stewart, *Real-World Haskell: Code You Can Believe In* (O'Reilly Media, 2008)

3. T. Sheard and S. P. Jones, "Template Metaprogramming for Haskell" in proceedings of the 2002 ACM SIGPLAN workshop on Haskell (ACM, 2002)

4. Case expressions and pattern matching, `https://www.cs.auckland.ac.nz/references/haskell/haskell-intro-html/patterns.html`

CHAPTER 12

Monads

Monads are important in Haskell, and they are used in many scenarios. The concept of monads can be confusing at first, but in this chapter, you will learn what monads are and how to use them in complex programs.

Introduction

A *monad* is a way for values to be used in sequences of computations, resulting in a structure of computations. The sequential building blocks can be used to create computations, and the building blocks themselves can be structured as computations. As the official Haskell documentation states, "It is useful to think of a monad as a strategy for combining computations into more complex computations."[1]

A monad is characterized by these three elements:

- A type constructor m (when working with monads, it is a good practice to name the type constructor m)

- A return function, which returns values of the type m

- A binding operation (>>=), which, by combining values of type m with computations that output values of type m, are used for producing new computations for m values

---

[1] https://wiki.haskell.org/All_About_Monads

© Stefania Loredana Nita and Marius Mihailescu 2019
S. L. Nita and M. Mihailescu, *Haskell Quick Syntax Reference*,
https://doi.org/10.1007/978-1-4842-4507-1_12

At the same time, `return` and `>>=` must follow *three laws*—right unit, left unit, and associativity. We will talk about these rules later in this chapter. A general representation of a monad is shown here:

```
data m a = ...
return :: a -> m a
(>>=)  :: m a -> (a -> m b) -> m b
```

On the first line, the type of the monad is m. On the second line, the value a is taken by the `return` function and is embedded into the monad m. On the third line, the binding function takes the monad instance m a and a computation that produces a monad instance m b from a's and produces the new monad instance m b.

One of the most common monads is the Maybe monad, whose type constructor m is Maybe. `return` and the binding operator have the following definition:

```
data Maybe a = Nothing | Just a

return :: a -> Maybe a
return x  = Just x

(>>=)  :: Maybe a -> (a -> Maybe b) -> Maybe b
m >>= g = case m of
                Nothing -> Nothing
                Just x  -> g x
```

Looking at the general structure of a monad and the definition of the Maybe monad, you can say the following about the Maybe monad: the type constructor is Maybe, and the `return` function takes a value and wraps it with Just, bringing the value into Maybe. The binding function takes as parameters the value m :: Maybe a and the function g :: a -> Maybe b. In other words, it looks like this: (a -> Maybe b) == g. It also shows how to work with g to m: if m is Nothing, then the result also will be Nothing, or g is applied to x, resulting in a value of Maybe b.

The Three Rules

You have seen that a monad must follow three rules: right unit, left unit, and associativity. These three rules show you the relation between a computation, the return function, and the binding operation.

All monads are instances of the Monad type class from Prelude, which is defined as follows:

```
class Monad m where
    return  :: a -> m a
    fail    :: String -> m a
    (>>=)   :: m a -> (a -> m b) -> m b
    (>>)    :: m a -> m b -> m b
```

In this definition, the last rule can be expressed in terms of the third rule. Note that (>>=) is read as "bind," and (>>) is read as "then."

Before going further, we need to mention that the do notation works great with monads; it acts as syntactic sugar for operations. You will see some examples in the next sections.

The Right Unit

This rule states the following:

```
f >>= return ≡ f
```

In this rule, f represents a computation. The rule says that if you make a computation f whose output is taken by return, then all that you did is nothing other than a computation. An example is the getLine function. The right unit rule applied on getLine states that reading a string and then returning the value is the same thing as just reading the string.

Using the do notation, the right unit rule says that the following programs make the same thing:

```
rightUnit1 = do
  x <- f
  return x

rightUnit2 = do
  f
```

The Left Unit

This rule states the following:

```
return a >>= f ≡ f a
```

In this rule, f is a computation, and a is a value. The rule says that if the result of a computation is a despite everything and you pass it to a computation f, then all what you did is to apply f directly on a. An example is the putStrLn function. The left unit rule applied on putStrLn says that if putStrLn takes a computation whose result is the value a, then this is the same thing as printing the value a.

Using the do notation, the right unit rule says that the following programs make the same thing:

```
leftUnit1 = do
  x <- return a
  f x

leftUnit2 = do
  f a
```

Associativity

The third rule says the following:

```
f >>= (\x -> g x >>= h) ≡ (f >>= g) >>= h
```

In this representation, the ≡ sign means "is equivalent." For short, this is the associativity rule for monads. To better understand, let's simplify the rule for the moment.

```
f >>= (g >>= h) ≡ (f >>= g) >>= h
```

In this version of the rule, it is pretty simple to identify the associativity. In other words, this means that when you make computations, it doesn't matter how they are grouped. Think of it as number addition: it doesn't matter how you group the numbers when you add them.

Next, let's take a look at (\x -> g x >>= h). This says that you take a value x, perform the computation g on the x, and send the result to h. The right side (f >>= g) >>= h says that the result of f is sent to g, and the result of g (performed on the result of f) is sent to h.

This a little complicated, but in a few words, the last rule says that when you have three computations, it doesn't matter the manner in which you group them because the result will be the same in all scenarios.

Using the do notation, the right unit rule says that the following programs make the same thing:

```
associativity1 = do
  x <- f
  do y <- g x
     h y

associativity2 = do
  y <- do x <- f
          g x
  h y
```

It is important to know that these three rules need to be assured by the programmer.

An Example

In this section, you will use the code examples provided at Yet Another Haskell Tutorial.[2]

Let's suppose you want to define a binary tree and then create a particular version of the map function that will apply a function to every leaf in the tree. It would look like this:

```
data Tree a
  = Leaf a
  | Branch (Tree a) (Tree a) deriving Show

mapTree :: (a -> b) -> Tree a -> Tree b
mapTree f (Leaf a) = Leaf (f a)
mapTree f (Branch lhs rhs) =
    Branch (mapTree f lhs) (mapTree f rhs)
```

This works just fine, but if you want to count the leaves from left to right, it would fail. To do this, you need a to use a *state*, which counts the leaves thus far. The function that uses states is called mapTreeState, and it is defined as follows:

```
mapTreeState :: (a -> state -> (state, b)) ->
                Tree a -> state -> (state, Tree b)
mapTreeState f (Leaf a) state =
    let (state', b) = f a state
    in  (state', Leaf b)
```

---

[2]https://en.wikibooks.org/wiki/Yet_Another_Haskell_Tutorial/Monads

```
mapTreeState f (Branch lhs rhs) state =
    let (state' , lhs') = mapTreeState f lhs state
        (state'', rhs') = mapTreeState f rhs state'
    in  (state'', Branch lhs' rhs')
```

The differences between mapTree and mapTreeState are that there are more arguments for f and the type -> Tree b was replaced with -> state -> (state, Tree b). To make it easier to work, let's use a type synonym declaration for the state, as shown here:

```
type State st a = st -> (st, a)
```

Next, add the following two functions that work on states:

```
returnState :: a -> State st a
returnState a = \st -> (st, a)

bindState :: State st a -> (a -> State st b) ->
             State st b
bindState m k = \st ->
    let (st', a) = m st
        m'       = k a
    in  m' st'
```

The function returnState keeps the state st and returns the value a. In other words, for the value a, it generates something of type State st a.

The function bindState transforms a into b. It works as follows: the initial state st is applied to m (whose type is State st a), resulting in a new state st and the value a. Next, applying the function k on a, it results in m' (whose type is State st b). Lastly, m' and the new state st' run.

Further, a new function is created, which uses the functions returnState and bindState.

```
mapTreeStateM :: (a -> State st b) -> Tree a -> State st (Tree b)
mapTreeStateM f (Leaf a) =
  f a `bindState` \b ->
  returnState (Leaf b)
mapTreeStateM f (Branch lhs rhs) =
  mapTreeStateM f lhs `bindState` \lhs' ->
  mapTreeStateM f rhs `bindState` \rhs' ->
  returnState (Branch lhs' rhs')
```

For a Leaf, the function f is applied to a, whose result is bound to a function that generates a Leaf with other value. For a Branch, the function begins with the left side and binds the result to a function that begins with the right side, whose result is bound with a function that generates a new Branch.

Having all these, you can say that State st is actually a monad with return implemented as returnState and (>>=) implemented as bindState. In [7], it is proved that these functions follow the three rules for monads.

To take it a step further, you can create the State st instance of Monad, but just writing instance Monad (State st) where { ... } won't work, because instances cannot be made from non-fully-applied type synonyms. To improve this, you can convert the type synonym to a newtype.

```
newtype State st a = State (st -> (st, a))
```

This implies that the State constructor needs to be packed and unpacked for the Monad instance declaration.

```
import Control.Applicative
import Control.Monad (liftM, ap)

instance Functor (State state)  where
  fmap = liftM
```

```
instance Applicative (State state)  where
  pure  = return
  (<*>) = ap
```

```
newtype State st a = State (st -> (st, a))
```

```
instance Monad (State state) where
    return a = State (\state -> (state, a))
    State run >>= action = State run'
        where run' st =
                    let (st', a)    = run st
                        State run" = action a
                    in  run'' st'
```

Putting it all together, the function mapTreeM looks like this:

```
mapTreeM :: (a -> State state b) -> Tree a ->
            State state (Tree b)
mapTreeM f (Leaf a) = do
  b <- f a
  return (Leaf b)
mapTreeM f (Branch lhs rhs) = do
  lhs' <- mapTreeM f lhs
  rhs' <- mapTreeM f rhs
  return (Branch lhs' rhs')
```

If the type signature is removed, then a more general version of mapTreeM is obtained.

```
mapTreeM :: Monad m => (a -> m b) -> Tree a ->
            m (Tree b)
```

In fact, mapTreeM can be applied in any monad, not just in State. Now, let's see some functions that take a current state and change it.

```
getState :: State state state
getState = State (\state -> (state, state))

putState :: state -> State state ()
putState new = State (\_ -> (new, ()))
```

The function getState returns the value of the current state, while putState inserts a new state, ignoring the current state.

Finally, to count the leaves, you can use the following function:

```
numberTree :: Tree a -> State Int (Tree (a, Int))
numberTree tree = mapTreeM number tree
    where number v = do
              cur <- getState
              putState (cur+1)
              return (v,cur)
```

To run the action, you should provide an initial state.

```
runStateM :: State state a -> state -> a
runStateM (State f) st = snd (f st)
```

Now let's put it all together. The final code should look like this:

```
import Control.Applicative
import Control.Monad (liftM, ap)

instance Functor (State state)  where
  fmap = liftM

instance Applicative (State state)  where
  pure  = return
  (<*>) = ap

data Tree a
  = Leaf a
  | Branch (Tree a) (Tree a) deriving Show
```

```haskell
newtype State st a = State (st -> (st, a))

instance Monad (State state) where
    return a = State (\state -> (state, a))
    State run >>= action = State run'
        where run' st =
                    let (st', a)    = run st
                        State run'' = action a
                    in  run'' st'

mapTreeM :: Monad m => (a -> m b) -> Tree a -> m (Tree b)
mapTreeM f (Leaf a) = do
 b <- f a
 return (Leaf b)
mapTreeM f (Branch lhs rhs) = do
 lhs' <- mapTreeM f lhs
 rhs' <- mapTreeM f rhs
 return (Branch lhs' rhs')

getState :: State state state
getState = State (\state -> (state, state))

putState :: state -> State state ()
putState new = State (\_ -> (new, ()))

numberTree :: Tree a -> State Int (Tree (a, Int))
numberTree tree = mapTreeM number tree
    where number v = do
                cur <- getState
                putState (cur+1)
                return (v,cur)
runStateM :: State state a -> state -> a
runStateM (State f) st = snd (f st)
```

Put it into a file called Tree.hs and then load it into GHCi (don't forget to change the current directory with the directory that contains the file Tree.hs).

```
Prelude> :load Tree.hs
[1 of 1] Compiling Main                    ( Tree.hs, interpreted )
Ok, one module loaded.
```

And now, let's see an example of tree.

```
testTree =
  Branch
    (Branch
      (Leaf 'a')
      (Branch
        (Leaf 'b')
        (Leaf 'c')))
    (Branch
      (Leaf 'd')
      (Leaf 'e'))
```

Let's apply the function runStateM and then print values in leaves.

```
*Main> runStateM (numberTree testTree) 1
Branch (Branch (Leaf ('a',1)) (Branch (Leaf ('b',2)) (Leaf
('c',3)))) (Branch (Leaf ('d',4)) (Leaf ('e',5)))

*Main> mapTreeM print testTree
'a'
'b'
'c'
'd'
'e'
Branch (Branch (Leaf ()) (Branch (Leaf ()) (Leaf ()))) (Branch
(Leaf ()) (Leaf ()))
```

Useful Combinators

In the Monad/Control.Monad, you can find some useful monadic combinators (note that m is an instance of Monad).

```
(=<<) :: (a -> m b) -> m a -> m b
mapM :: (a -> m b) -> [a] -> m [b]
mapM_ :: (a -> m b) -> [a] -> m ()
filterM :: (a -> m Bool) -> [a] -> m [a]
foldM :: (a -> b -> m a) -> a -> [b] -> m a
sequence :: [m a] -> m [a]
sequence_ :: [m a] -> m ()
liftM :: (a -> b) -> m a -> m b
when :: Bool -> m () -> m ()
join :: m (m a) -> m a
```

We will not discuss these combinators here, so please check the documentation.[3]

Summary

In this chapter, you learned about monads.

- What a monad is

- What the rules are that structures and computations need to follow to be considered a monad

- How to create a monad

- What are some useful combinators

As a final remark in this chapter, monads are extremely important in Haskell, so you need to understand them well.

---

[3]http://hackage.haskell.org/package/base-4.12.0.0/docs/Control-Monad.html

References

1. C. A. R. Hoareetal, "Tackling the Awkward Squad:
    Monadic Input/Output, Concurrency, Exceptions,
    and Foreign-Language Calls in Haskell," *Engineering
    Theories of Software Construction*" (2001)

2. M. Maruseac, A. S. Mena, A. Abel, A. Granin,
    H. Apfelmus, D. Austin, and J. Breitner, "Haskell
    Communities and Activities Report" (2017)

3. M. P. Jones and L. Duponcheel, "Composing
    Monads," Technical Report YALEU/DCS/
    RR-1004 (Department of Computer Science,
    Yale University, 1993)

4. T. Schrijvers, and B. CdS Oliveira, "Monads, Zippers,
    and Views: Virtualizing the Monad Stack," ACM
    SIGPLAN Notices 46.9: 32–44 (2011)

5. J. Hedges, "Monad Transformers for Backtracking
    Search," arXiv preprint arXiv:1406.2058 (2014)

6. All about monads, `https://wiki.haskell.org/`
    `All_About_Monads`

7. Yet another Haskell tutorial/monads, `https://`
    `en.wikibooks.org/wiki/Yet_Another_Haskell_`
    `Tutorial/Monads`

8. `Control.Monad`, `http://hackage.haskell.org/`
    `package/base-4.12.0.0/docs/Control-Monad.html`

CHAPTER 13

Monad Transformers

In the previous chapter, you saw how useful monads are. But what if you need operations from two different monads? In this chapter, you will learn how to proceed in such scenarios.

Simple Transformers

We'll start with a brief description of transformers: a *transformer* represents a special type that allows you to combine two monads into one monad that uses the properties of both monads. The result of combining two monads is also a monad.

In Haskell, there are two main packages related to monad transformers.

- `transformers`, which includes the monad transformer class and actual transformers. The most commonly used classes in this package are `MonadTrans` and `MonadIO`. All transformers in this package are instances of `MonadTrans`, and they are used to create new monads from existing monads.

- `mtl` (Monad Transform Library), which includes instances of different monad transformers, based on functional dependencies.

© Stefania Loredana Nita and Marius Mihailescu 2019
S. L. Nita and M. Mihailescu, *Haskell Quick Syntax Reference*,
https://doi.org/10.1007/978-1-4842-4507-1_13

These packages are compatible for use together; they have in common classes, type constructors, and functions, although the names of the modules are different. The transformers package is independent of functional dependencies, being more portable than mtl. Still, the operations in one monad need to be lifted by the programmer in the resulting monad since the actual transformers do not have monad classes.

When you create a monad transformer, the convention is that the resulting monad keeps the name of the base monad, followed by the T character. For example, the transformer for the Maybe monad is MaybeT.

The following are some examples of transformers (m in the following examples represents an arbitrary monad):

- MaybeT: A monad incorporates Maybe a, resulting in m (Maybe a).

- ReaderT: The result a in Reader r a is incorporated by another monad, resulting in r-> m a.

- StateT: From State s, a monad incorporates the return value and its state (a,s), resulting in s -> m (a,s).

- ExceptT: A monad incorporates Either e a, resulting in m (Either e a).

MaybeT Transformer

In this section, you will see how the MaybeT transformer was obtained from the Maybe monad. The code in this section is based on [6].

The first step is to define a new type.

```
newtype MaybeT m a = MaybeT { runMaybeT :: m (Maybe a) }
```

Here, the type constructor is MaybeT, with the parameter m and a term constructor of MaybeT. You can access this representation through the function runMaybeT.

You saw in the previous chapter that all monads are instances of the Monad class, so you need to make MaybeT an instance of Monad. Remember that a monad needs to contain the return function and the binding operation (>>=).

```
instance Monad m => Monad (MaybeT m) where
  return  = MaybeT . return . Just

-- (>>=) :: MaybeT m a -> (a -> MaybeT m b) -> MaybeT m b
  x >>= f = MaybeT $ do maybe_value <- runMaybeT x
                        case maybe_value of
                            Nothing    -> return Nothing
                            Just value -> runMaybeT $ f value
```

Let's take a look at the binding operation, which wraps and unwraps a few times. In the first instruction from the do block into the runMaybeT, the value x is unwrapped into the m (Maybe a) computation, from which Maybe a are extracted through ->. In the case statement, maybe_value is tested, returning Nothing into m in the case of Nothing, while, in the case of Just, f is applied on the value of Just. In addition, in the Just case, runMaybeT puts the output into monad m, because the result of f has type MaybeT m b. The type of the entire do block is m (Maybe b), which is wrapped into the MaybeT constructor.

The function maybe_value is defined in terms of the bind operator from Maybe.

```
maybe_value >>= f = case maybe_value of
                        Nothing -> Nothing
                        Just value -> f value
```

Although the runMaybeT is in do block, you need to use the MaybeT constructor before do, because the do block needs to be in m, not in MaybeT m.

Next, MonadT m needs to be set as an instance of Monad, Applicative, and Functor.

```
instance Monad m => Applicative (MaybeT m) where
    pure = return
    (<*>) = ap
```

```
instance Monad m => Functor (MaybeT m) where
    fmap = liftM
```

It is natural to make a MaybeT m instance of MonadTrans, because all transformers are instances of MonadTrans. They are all instances of Alternative and MonadPlus as well, because Maybe is an instance of these two.

```
import Control.Applicative
import Control.Monad (MonadPlus, liftM, ap)
import Control.Monad.Trans
import Control.Monad (mplus, mzero, liftM, ap)
```

```
instance Monad m => Alternative (MaybeT m) where
    empty   = MaybeT $ return Nothing
    x <|> y = MaybeT $ do maybe_value <- runMaybeT x
                          case maybe_value of
                              Nothing   -> runMaybeT y
                              Just _    -> return maybe_value
```

```
instance Monad m => MonadPlus (MaybeT m) where
    mzero = empty
    mplus = (<|>)
```

```
instance MonadTrans MaybeT where
    lift = MaybeT . (liftM Just)
```

The function `lift` from `MonadTrans` takes functions from the monad `m`, bringing them into the `MaybeT m` monad. In this way, they can be used in do blocks.

Building a Simple Monad Transformer Stack

In this section, you will use the following monads in this order to build a monad transformer stack: `Identity`,[1] `Reader`,[2] and `Writer`.[3] The `Identity` monad does not do anything important, but it is useful in certain situations. The `Reader` monad can be used when you want to add information to a pure function. Lastly, the `Writer` monad can be used when you want to add logging to a function.

Another thing you need is the `lift`[4] function from `Control.Monad. Trans.Class`, which takes a computation from a monad and allows you to use that computation in another monad. It is useful when you want to work with the `ask` function from `Reader`, which is contained in a transformer stack.

Let's write the example.

```
import Control.Monad.Trans.Class (lift)
import Data.Functor.Identity (Identity, runIdentity)
import Control.Monad.Trans.Reader (ReaderT, ask, runReader)
import Control.Monad.Trans.Writer (WriterT, tell, runWriter)
```

---

[1]https://hackage.haskell.org/package/base-4.8.0.0/docs/Data-Functor-
Identity.html

[2]https://hackage.haskell.org/package/transformers-0.4.3.0/docs/
Control-Monad-Trans-Reader.html

[3]https://hackage.haskell.org/package/transformers-0.4.3.0/docs/
Control-Monad-Trans-Writer-Lazy.html

[4]https://hackage.haskell.org/package/transformers-0.4.3.0/docs/
Control-Monad-Trans-Class.html#v:lift

```
type DataIn = Integer
type DataOut = [String]
type Outcome = Integer

transformersStack :: WriterT DataOut (ReaderT DataIn Identity)
Outcome
transformersStack = do
    y <- lift ask
    tell ["The user introduced: " ++ show y]
    return y
```

To test this, you need to import Control.Monad.Trans.Writer
and Control.Monad.Trans.Reader and then load your file (called
TransformersStack.hs).

```
Prelude> import Control.Monad.Trans.Writer
Prelude Control.Monad.Trans.Writer> import Control.Monad.Trans.
Reader
Prelude Control.Monad.Trans.Writer Control.Monad.Trans.Reader>
:load TransformersStack.hs
[1 of 1] Compiling Main      ( TransformersStack.hs, interpreted )
Ok, one module loaded.
*Main Control.Monad.Trans.Writer Control.Monad.Trans.Reader>
let myReader = runWriterT transformersStack
*Main Control.Monad.Trans.Writer Control.Monad.Trans.Reader>
let myIdentity = runReaderT myReader 7
*Main Control.Monad.Trans.Writer Control.Monad.Trans.Reader>
runIdentity myIdentity
(7,["The user introduced: 7"])
```

The Identity monad is a base monad that lets you put things on the
top (the other base monad is IO). Before the Identity monad is the Reader
monad, which holds a number in the previous example. At the top of the
transformers stack is the Write monad, which takes a list of strings.

108

The external `Writer` monad could be used directly with `tell`, but the calls needs to be wrapped into the internal `Reader`. This is done by the `lift` function. Note that the `lift` function is useful in the monad transformer stack no matter the number of monads.

Summary

In this chapter, you learned what monad transformers are and how they can be used. You saw a simple example of building a monad transformer. Finally, you learned how to create a stack of monad transformers.

References

1. M. P. Jones, "Functional Programming with Overloading and Higher-Order Polymorphism," International School on Advanced Functional Programming(Springer, 1995)

2. Transformers: concrete functor and monad transformers, `http://hackage.haskell.org/package/transformers`

3. `mtl`: monad classes, using functional dependencies, `http://hackage.haskell.org/package/mtl`

4. Monad transformers, `https://wiki.haskell.org/Monad_Transformers`

5. Monad transformers, `https://www.schoolofhaskell.com/user/commercial/content/monad-transformers`

6. All about monads, `https://wiki.haskell.org/`
    `All_About_Monads`

7. Haskell/monad transformers, `https://`
    `en.wikibooks.org/wiki/Haskell/Monad_`
    `transformers`

8. All about monads, `https://wiki.haskell.org/`
    `All_About_Monads#Monad_transformers`

CHAPTER 14

Parsec

Now that you know how to work with monads, let's take a step further. In this chapter, you will learn how to use the parsec library.

The parsec library is an "industrial-strength, monadic parser combinator library for Haskell."[1] It's used to create larger expressions by constructing parsers through a combination of different higher-order combinators. The main module of parsec is Text.Parsec, which provides ways to parse to Char data. The other module of parsec is Text.ParserCombinators.Parsec.

Parsec is useful when you are working with many user inputs and the users need to understand the error messages. The library may take different types of inputs, for example, Strings or tokens (strings with an assigned meaning) provided by an external program that performs lexical analysis (regarding Text.Parsec). The library also can be layered into a monad stack through the monad transformer that it provides (Text.ParserCombinators.Parsec). The monad transformer is useful in scenarios in which additional states of parsing need to be monitored.

The parsec package should be installed by default, but if it is not installed, open a terminal (if you use a Windows operating system, it is recommended that you choose the option Run as Administrator) and then type the following:

```
cabal install parsec
```

---

[1]Parsec, https://wiki.haskell.org/Parsec

© Stefania Loredana Nita and Marius Mihailescu 2019
S. L. Nita and M. Mihailescu, *Haskell Quick Syntax Reference*,
https://doi.org/10.1007/978-1-4842-4507-1_14

For the moment, take this instruction as it is; you will learn about cabal in Chapter 26.

Now let's take a look at the example given on the official page of the package:[2]

```
Prelude> import Text.Parsec
Prelude Text.Parsec> let parenSet = char '(' >> many parenSet
>> char ')' :: Parsec String () Char
Prelude Text.Parsec> let parens = (many parenSet >> eof) <|> eof
Prelude Text.Parsec> parse parens "" "()"
Right ()
Prelude Text.Parsec> parse parens "" "()(())"
Right ()
Prelude Text.Parsec> parse parens "" "("
Left (line 1, column 2):
unexpected end of input
expecting "(" or ")"
```

This example creates a parser that verifies whether the input contains matching parentheses.

Let's analyze the functions that are being used. The function char takes as input a single character and returns the parsed character. Next (>>) is a function from the Monad class, which takes a constant function, which will be mapped over an instance of a monad and finally flatten the result (note that this function needs to return an instance of the monad itself). The function many applies a parser zero or more times and returns a list of the results of the parser. The eof parser succeeds only at the end of the input. Finally, the combinator (<|>) takes two parsers: if the left parser succeeds, then it applies this parser; otherwise, it tries the right parser.

---

[2]http://hackage.haskell.org/package/parsec

You get two results: Right() indicates that there are matching parentheses, while Left... indicates a failure, detailing the error.

You can find some great documentation of parsec, including comprehensive examples, in [2]. Other relevant examples are in Chapter 16 of [3].

Summary

In this chapter, you learned what parsec is and when you can use it. Also, you examined a short example of using the functions of this package.

References

1.   D. Leijen and E. Meijer. "Parsec: Direct Style Monadic Parser Combinators for the Real World," (2001)

2.   Parsec, a fast combinator parser, https://web.archive.org/web/20140528151730/http://legacy.cs.uu.nl/daan/parsec.html

3.   B. O'Sullivan, J. Goerzen, and D. B. Stewart, *Real-World Haskell: Code You Can Believe In* (O'Reilly Media, 2008)

4.   Parsec, https://wiki.haskell.org/Parsec

5.   parsec: monadic parser combinators, http://hackage.haskell.org/package/parsec

CHAPTER 15

Folds

An important element in functional programming is the *fold* (also known as *reduce*). A fold represents a set of higher-order functions that operate on recursive data structures and take a combining operation as one of the parameters, recombining the results of recursive operations in order to build a return value. In this chapter, you will learn to use fold functions.

The fold family contains the following functions:

- `fold` takes as one of the parameters a function and *folds* it onto a list.

- `foldr` takes as one of the parameters a function and *folds it from right to left* onto a list.

- `foldl` takes as one of the parameters a function and *folds it from right to left* onto a list.

- `foldr1` is similar to `foldr`.

- `foldl1` is similar to `foldl`.

To better understand these functions, let's examine the following example:

```
Prelude> foldl (+) 5 [1..6]
26
```

© Stefania Loredana Nita and Marius Mihailescu 2019
S. L. Nita and M. Mihailescu, *Haskell Quick Syntax Reference*,
https://doi.org/10.1007/978-1-4842-4507-1_15

In the first step, you have a list.

		1	2	3	4	5	6

Then the function that is the first argument is folded into the elements.

	+	1	+	2	+	3	+	4	+	5	+	6

Next, foldl needs a starting point, which is taken from the second argument (note there is nothing in the leftmost piece).

5	+	1	+	2	+	3	+	4	+	5	+	6

Indeed, 5+1+2+3+4+5+6=26; therefore, you get the result 26.

Note that you used the left fold, which is the reason why you started folding from the left. Another important note is that the fold functions are not limited to infix functions as the first argument.

Further, let's take a look at foldr and foldr1. The definition of foldr is as follows:

```
foldr  :: (a -> b -> b) -> b -> [a] -> b
```

The definition of foldr1 is as follows:

```
foldr1 :: (a -> a -> a) ->      [a] -> a
```

From here, you can see that both of them have as the first argument a function that takes two arguments. The differences are as follows:

- For foldr, the two arguments may have different types, resulting in output with the same type as the second argument.

- For foldr1, the two arguments must have the same type, resulting in output of the same type as the arguments. Another aspect is that it does not need the starting point (it will choose the rightmost element to start with).

Similar to the fold functions are scan functions, which contain the following: scanl, scanr, sccanl1, and scanr1. Basically, these are analogous to foldl, foldr, foldl1, and foldl2, but they show the intermediate results. Returning to the example, if you apply scanl instead of foldl, you will get the following:

```
Prelude> scanl (+) 5 [1..6]
[5,6,8,11,15,20,26]
```

The opposite function of foldr is unfoldr. While foldr reduces a list to just one value, unfoldr builds a list based on a seed. Its definition is as follows:

```
unfoldr :: (b -> Maybe (a, b)) -> b -> [a]
```

Observe that unfoldr returns Nothing if it finished building or Just(a,b) otherwise, where a represents the list and b represents the next element. Let's see an example[1]:

```
Prelude Data.List> unfoldr (\b -> if b == 0 then Nothing else
Just (b, b-1)) 10
[10,9,8,7,6,5,4,3,2,1]
```

Summary

In this chapter, you learned the following:

- What fold family functions are and how to work with them

- What the opposite of folding is

---

[1]The example is taken from http://hackage.haskell.org/package/base-4.12.0.0/docs/Data-List.html.

Reference

1. Data.List, http://hackage.haskell.org/
    package/base-4.12.0.0/docs/Data-List.html

CHAPTER 16

Algorithms

In this chapter, you'll learn how to implement algorithms such as quicksort, mergesort, and bubble sort.

Quicksort

Quicksort is a divide-and-conquer algorithm. Quicksort divides a large array into two smaller subarrays, known as the *low elements* and *high elements*. The time complexity in the best case is $O(n \log n)$ and in the worst case is $O(n^2)$.

The steps are as follows:

1. Choose an element from the array. The element will be called the *pivot*.

2. Reorder the array in such a way that all the elements with values less than the pivot come before the pivot, while the elements with values that are greater than the pivot come after it. After the partitioning process, the pivot is in its final position. This is called the *partition operation*. The equal values don't matter; they can go on any branch.

© Stefania Loredana Nita and Marius Mihailescu 2019
S. L. Nita and M. Mihailescu, *Haskell Quick Syntax Reference*,
https://doi.org/10.1007/978-1-4842-4507-1_16

3.   Recursively you will need to apply these steps to
the subarray elements that have smaller values and
separately to the subarray of elements with values
that are greater.

The function qs (quicksort) will take a list [a] of some type a, such that
elements of a can be compared with each other (this is specified using the
(Ord a) guard). Then, the function will return a list of the same type [a].

```
qs :: (Ord a) => [a] -> [a]
```

The quicksort is implemented in Haskell as follows:

```
qs (y:ys) = qs [x | x <- ys, x<= y] ++ [y] ++ qs [x | x <- ys, x > y]
```

qs takes as an argument (y:ys), which is a list consisting of the first
element, x, and the rest of the list, ys. You apply the list comprehension
[x | x <-ys, x <= y] to get a list of all the elements in the list ys that
are smaller or equal than y. Next, you concatenate the resulting list with a
single element list, [y], and the list of elements that are greater than x.

The recursion in the quicksort algorithm is defined by the function qs,
but you still need to finish somehow the recursion at a certain point, so
you need to specify the condition when the recursion will end. This is also
easily done in Haskell by augmenting the definition of the function qs by
adding one more extra rule.

```
qs [] = []
```

The algorithms applied on an empty list will return an empty list.

By combining everything, you get the complete quicksort
implementation in Haskell, as shown here:

```
qs :: (Ord a) => [a] -> [a]
qs [] = []
qs (y:ys) = qs [x | x <- ys, x<= y] ++ [y] ++ -qs[x | x <- ys, x > y]
```

```
Prelude> qs [1,2,7,8,9,5,3,3]
[1,2,3,3,5,7,8,9]
```

Mergesort

Compared with quicksort, mergesort is a little more complicated during the implementation process. The time complexity is O(n log n).
The algorithm steps are as follows:

1. The list is divided into two parts.

2. The two parts are sorted by the algorithm.

3. The sorted parts are merged by a special merging procedure that is dedicated to sorted lists.

Here is the procedure for splitting the list into two parts:

```
ms'splitinhalf :: [a] -> ([a], [a])
ms'splitinhalf ys = (take m ys, drop m ys)
      where m = (length ys) `div` 2
```

Let's analyze this code. The function ms'splitinhalf will return a pair of arrays into which the original array was divided into two. The m is equal to half of the length of the array, and the standard functions take and drop are used to get the first m elements of the list take m ys and the rest of the elements of the list after those first elements that are used with drop m ys.

Next, you have to define the function that merges the two sorted arrays.

```
ms'splitinhalf :: [a] -> ([a], [a])
ms'splitinhalf ys = (take m ys, drop m ys)
      where m = (length ys) `div` 2

ms'merge :: (Ord b) => [b] -> [b] -> [b]
ms'merge [] ys = ys
ms'merge ys [] = ys
```

```
ms'merge (y:ys) (x:xs)
      | (y < x) = y:ms'merge ys (x:xs)
      | otherwise = y:ms'merge (y:ys) ys
```

The function will receive two arrays and will produce one array of the same type. The algorithm for merging is as follows:

1. If the first list is empty, [ ], the result of the merge function is the second list, ys.

2. If the second list is empty, [ ], then the result of the merge is the first list, xs.

3. Otherwise, you will return the first elements of the lists and append with the colon (:) function the least of them to the new list, which is the result of merging the two remaining lists.

After you have defined the functions ms'splitinhalf and ms'merge, you can easily define the function mergesort.

```
ms :: (Ord a) => [a] -> [a]
ms xs
      | (length xs) > 1 = ms'merge (ms ls) (ms rs)
      | otherwise = xs
    where (ls, rs) = ms'splitinhalf xs
```

Now, if the length of the list is bigger than 1, then you follow the default steps of the algorithm. Otherwise, the list with a length of 1 will already be sorted, which represents the condition for ending the recursion.

The complete code for the mergesort is shown here:

```
ms'merge :: (Ord a) => [a] -> [a] -> [a]
ms'merge [] xs = xs
```

```
ms'merge xs [] = xs
ms'merge (x:xs) (y:ys)
    | (x < y) = x:ms'merge xs (y:ys)
    | otherwise = y:ms'merge (x:xs) ys

ms'splitinhalf :: [a] -> ([a], [a])
ms'splitinhalf xs = (take n xs, drop n xs)
    where n = (length xs) 'div' 2

ms :: (Ord a) => [a] -> [a]
ms xs
    | (length xs) > 1 = ms'merge (ms ls) (ms rs)
    | otherwise = xs
    where (ls, rs) = ms'splitinhalf xs

Prelude> ms [1,3,4,1,2,23,7,8,5]
[1,1,2,3,4,5,7,8,23]
```

Bubble sort

For a bubble sort operation, you change the placement of the element pairs, while there is still a pair of elements (x,y) such as x > y. The time complexity for the worst case is $O(n^2)$, and for the best case it is $O(n)$.

You will define a function that will iterate through all the elements in a list, and it will switch the pairs of the unsorted elements.

```
bs'iter :: (Ord a) => [a] -> [a]
bs'iter (x:y:xs)
    | x > y = y : bsiter (x:xs)
    | otherwise = x : bs'iter (y:xs)
bs'iter (x) = (x)
Prelude> bs'iter [2,1,3]
[3,2,1]
```

Next, you just need to apply the next function n times. The function represents the length of the list that should be sorted.

```
bs :: (Ord a) => [a] -> Int -> [a]
bs xs i
    | i == (length xs) = xs
    | otherwise = bs' (bs'iter xs) (i + 1)

bs :: (Ord a) => [a] -> [a]
bs xs = bs' xs 0

Prelude> bs' [3,2,1,5] 4
[3,2,1,5]
```

You can do this by defining the function bs', which will take two arguments: the list and the number of the current iteration, which is i.

This transforms the iteration into a recursion in such a way that bubble sorting will become a recursive algorithm (just for this example, because the bubble sort is not recursive like the original definition).

Summary

In this chapter, we discussed the three most important algorithms used for sorting: quicksort, mergesort, and bubble sort. We presented their implementations by providing the shortest and most flexible versions of how they can be implemented.

Reference

1. Cormen, Thomas H., Charles E. Leiserson, Ronald L. Rivest, and Clifford Stein. Introduction to algorithms. MIT press, 2009.

CHAPTER 17

Parsing

Wikipedia defines the parsing process as follows:

> *"The process of analyzing a string of symbols, either in natural language, computer languages, or data structures, conforming to the rules of a formal grammar. The term is used to refer to the formal analysis by a computer of a sentence or other string of words into its constituents, resulting in a parse tree showing their syntactic relation to each other, which may also contain semantic and other information."*

This sounds a little complex, and it really is, involving grammars, parsing trees, some tokens, expressions, terms, and so on. In this chapter, you will learn something easier: how to parse binary data in Haskell.

This chapter is inspired from [2] and will use the examples provided there. Note that the examples may not work with newer versions of ghc. To run them, use ghc 8.0.2.

When you have a string of bytes, you will want to do something with them. A package that deals with binary data in Haskell is called Binary. First you need to install it, so open a terminal and type the following:

```
cabal install binary
```

The Binary package has three main components.

- The Get monad
- The Put monad
- A serialization component

© Stefania Loredana Nita and Marius Mihailescu 2019
S. L. Nita and M. Mihailescu, *Haskell Quick Syntax Reference*,
https://doi.org/10.1007/978-1-4842-4507-1_17

Get is a state monad, so it keeps a state and changes it when an action is applied on that state. In this situation, the state is actually an offset of the lazy string of bytes that will be parsed. In scenarios where you get a failure when trying to parse a bytestring, the failure will be interpreted as an exception—these are handled in the IO monad due to laziness—which will probably be thrown in a different place than you expect. In such situations, you can use a stricter version of the Get monad.

Here is an example with a regular Get:

```haskell
import qualified Data.ByteString.Lazy as BL
import Data.Binary.Get
import Data.Word

deserialiseHeader :: Get (Word32, Word32, Word32)
deserialiseHeader = do
  alen <- getWord32be
  plen <- getWord32be
  chksum <- getWord32be
  return (alen, plen, chksum)

main :: IO ()
main = do
  input <- BL.getContents
  print $ runGet deserialiseHeader input
```

The inputs are three 32-bit numbers in big-endian format, and the output is a tuple.

```
% runhaskell /ch17/parsing.hs << EOF
heredoc> 123412341235
heredoc> EOF
(825373492,825373492,825373493)
```

If the input is too short, you get an exception.

```
% runhaskell / ch17/parsing.hs << EOF
tooshort
EOF
parsing.hs: Data.Binary.Get.runGet at position 8: not enough bytes
CallStack (from HasCallStack):
  error, called at libraries/binary/src/Data/Binary/Get.
  hs:351:5 in binary-0.8.6.0:Data.Binary.Get
```

In the previous example, the function getWord32 takes the input, goes through it, and returns a value.

The next example decodes a list of numbers that end with EOF in a recursive manner:

```
listOfWord16 = do
  empty <- isEmpty
  if empty
    then return []
    else do v <- getWord64be
            rest <- listOfWord16
            return (v : rest)
```

You saw in the first example that you get an exception when the input is too short. There are two ways to handle exceptions: write your own parser or handle the exceptions in the IO monad. The second way is simpler because it involves a stricter version of the Get monad, in which the parser Get a taken as input to runGet results in (Either String a, ByteString). This means the first value is either a string (i.e., the exception message) or the result, while the second value is the remaining bytestring.

The following is the modified version of the first example with the strict Get being used:

```
import qualified Data.ByteString as B
import Data.Binary.Strict.Get
import Data.Word

deserialiseHeader :: Get (Word32, Word32, Word32)
deserialiseHeader = do
  alen <- getWord32be
  plen <- getWord32be
  chksum <- getWord32be
  return (alen, plen, chksum)

main :: IO ()
main = do
  input <- B.getContents
  print $ runGet deserialiseHeader input
```

Note that this example requires binary-strict, which needs to be installed using cabal. The changes in the code are that it is using a strict bytestring instead of a lazy bytestring and it is importing Data.Binary. Strict.Get. If the example is run again, you will obtain:

```
% runhaskell /ch17/parsing.hs << EOF
heredoc> 123412341235
heredoc> EOF
(Right (825373492,825373492,825373493),"\n")
```

Now it works correctly because the output is a Right, and a new line was added instead of being consumed by the parser. Let's run it with the shorter input, as shown here:

```
% runhaskell /ch17/parsing.hs << EOF
heredoc> tooshort
heredoc> EOF
(Left "too few bytes","\n")
```

Now, the output is a Left instead of an exception and can be handled in the IO monad. The fail can be called inside the parser.

Operations on bits are allowed. Importing Data.Bits, the following operators can be used:

Operator	Symbol
AND	.&.
OR	.|.
XOR	`xor`
NOT	`complement`
Left shift	`shiftL`
Right shift	`shiftR`

Don't forget that working with bits needs special attention.

Summary

In this chapter, you learned how to parse a bytestring using binary and binary-strict.

For a parser created from scratch, check out [3].

References

1. G. Hutton and E. Meijer, "Monadic Parsing in Haskell," *Journal of Functional Programming*, 8.4: 437–444 (1998)

2. Dealing with binary data, `https://wiki.haskell.org/Dealing_with_binary_data#Binary_parsing`

3. Parser, `https://www.schoolofhaskell.com/school/starting-with-haskell/basics-of-haskell/8_Parser`

CHAPTER 18

Parallelism and Concurrency

Parallelism is a computing strategy that enables many computations (or the execution of processes) to be performed simultaneously. In this chapter, you will learn the basic elements of parallelism and concurrency in Haskell.

Before continuing, let's see what the differences are between parallelism and concurrency. In parallel computing, a larger problem is divided into smaller ones that are solved simultaneously, which implies that the hardware needs to have many processing units. Mainly, the purpose of parallelism is to make programs faster by adopting a strategy in which the dependencies between data are at a minimum. By contrast, the purpose of concurrency is to make programs more usable, and it can be used on a single processing unit (although it is compatible with multiple processing units to increase the speed). In concurrency, a problem can be executed partially without affecting the final output. When you are dealing with concurrency, it involves distributed computing.

Typically, in parallelism, you work with *processes* (do not confuse *processes* with *processors*, which refers to the hardware component), while in concurrency you work with *threads*. As concurrency can be executed on a single processing unit, you can easily deduce that a process contains at least one thread.

© Stefania Loredana Nita and Marius Mihailescu 2019
S. L. Nita and M. Mihailescu, *Haskell Quick Syntax Reference*,
https://doi.org/10.1007/978-1-4842-4507-1_18

Note that parallelism and concurrency are closely related and are used in many situations together.

Parallelism

In Haskell, there are two ways to use parallelism.

- Using `Control.Parallel`, which leads to pure parallelism

- Using concurrency to parallelize IO

The advantages of using parallel programming are that you obtain the same result every time (i.e., determinism) and you don't get race conditions (i.e., the system depends on the timing or sequences of uncontrollable events) or deadlocks (a number of processes/threads expects a response from another process/thread).

The `Control.Concurrent` module is in the `parallel` package[1] and contains the following two combinators:

```
infixr 0 `par`
infixr 1 `pseq`

par  :: a -> b -> b
pseq :: a -> b -> b
```

The `par` combinator says that the first argument is evaluated at the same time as the second argument, but it returns the value of the second argument.

For these scenarios, you'll meet a new term, *spark*. So, in the x `par` y statement, the evaluation of x is sparked, but y is returned. The sparks are stored in a queue and are executed in first in, first out (FIFO) order, not immediately. If at the time of an execution step an idle central unit

---

[1]http://hackage.haskell.org/package/parallel

processing (CPU) is detected, then a spark is converted into a thread that will run on the idle CPU.

The combinator pseq is similar to seq,[2] but the difference is at runtime: seq may evaluate the arguments in any order, and pseq is forced to evaluate the first argument before the second one.

A simple example of parallelism is the Fibonacci example.[3]

```
import Control.Parallel

nfib :: Int -> Int
nfib n | n <= 1 = 1
       | otherwise = par n1 (pseq n2 (n1 + n2 + 1))
                     where n1 = nfib (n-1)
                           n2 = nfib (n-2)
```

Let's look for values where n>1. Here, par sparks the thread to evaluate nfib(n-1), while pseq forces the evaluation of nfib(n-2) on the parent thread. Through pseq, the nfib(n-2) branch is evaluated before addition with nfib(n-1). This approach is actually a divide-and-conquer operation, in which the parent thread evaluates a branch, while a new thread is sparked to evaluate the other branch. The combinatory pseq ensures that n2 is evaluated before n1 in the parent thread in the expression (n1+n2+1). This is mandatory because the compiler may not generate code to evaluate the addends from left to right.

More complex combinators are provided in the Control.Parallel. Strategies[4] module. Here, the operations are created around the par combinator, which provides more complex patterns for parallel computing.

---

[2]http://hackage.haskell.org/package/base-4.11.1.0/docs/Prelude.html#v:seq

[3]Here's an example taken from some older Haskell documentation: https://downloads.haskell.org/~ghc/8.6.3/docs/html/users_guide/parallel.html#parallel-haskell

[4]https://downloads.haskell.org/~ghc/6.6.1/docs/html/libraries/base/Control-Parallel-Strategies.html

Concurrency

In Haskell, concurrency (provided through `Control.Concurrent`) is accomplished by using threads from the monad IO. One of the greatest models in concurrency is software transactional memory (STM), which works with `forkIO` and `MVars`. We will not cover parallel and concurrent programming in this chapter because they require a whole separate discussion. In [8] you will find a great list of documentation about concurrent and parallel programming in Haskell.

Summary

In this chapter, you learned the following:

- What parallelism and concurrency are

- The difference between parallelism and concurrency

- How the function `Control.Parallel` can be used

References

1. S. P. Jones, A. Gordon, and S. Finne, "Concurrent Haskell," *POPL*, vol. 96 (1996)

2. S. Marlow, *Parallel and Concurrent Programming in Haskell: Techniques for Multicore and Multithreaded Programming* (O'Reilly Media, 2013)

3. S. P. Jones and S. Singh, "A Tutorial on Parallel and Concurrent Programming in Haskell," International School on Advanced Functional Programming (Springer, 2008)

4. parallel: Parallel programming library, `http://hackage.haskell.org/package/parallel`

5. Prelude, `http://hackage.haskell.org/package/base-4.11.1.0/docs/Prelude.html`

6. Chapter 7, GHC language features, `https://downloads.haskell.org/~ghc/7.0.3/docs/html/users_guide/lang-parallel.html`

7. `Control.Parallel.Strategies`, `https://downloads.haskell.org/~ghc/6.6.1/docs/html/libraries/base/Control-Parallel-Strategies.html`

8. Parallel/reading, `https://wiki.haskell.org/Parallel/Reading`

CHAPTER 19

Haskell Pipes

Haskell streaming programs provide great features such as effects, streaming, and composability. In classical programming, you can choose just two of these features. In this chapter, you will learn how to use pipes to achieve all three.

If you renounce *effects*, then you will obtain lists that are pure and lazy (we will talk about laziness in Haskell in Chapter 21). You will be able to transform them by applying composable functions in constant space, but you will not be able to interleave the effects. If you renounce *streaming*, then you will obtain mapM (which maps every element of a structure to a monadic action, and after the actions are evaluated from left to right, the results are collected), forM (similar to mapM but with flipped arguments), and a version of ListT that will not work properly. These imply effects and composability, but the result is returned only after the whole list is processed and loaded into memory. Lastly, if you renounce *composability*, then you will be able to write dependent reads, writes, and transforms, but they won't be separate or modular.

A way to get all three functionalities is to use pipes, provided by the pipes[1] library. This library provides abstractions such as Producer, Consumer, Pipe, and the correct version of ListT, which can be combined in any way, because they have the same base type.

---

[1]http://hackage.haskell.org/package/pipes

© Stefania Loredana Nita and Marius Mihailescu 2019
S. L. Nita and M. Mihailescu, *Haskell Quick Syntax Reference*,
https://doi.org/10.1007/978-1-4842-4507-1_19

Specifically, with pipes, levels of streaming processing are forced to be decomposed such that they can be combined. This approach is useful because streaming components can be reused as interfaces or can be connected using constant memory if they are premade.

To decouple data, there are two commands: yield, which sends output data, and await, which receives input data. The following are the monad transformers and the contexts in which they can be used:

- Producer is used only with yield and models streaming sources.

- Consumer is used only with await and models streaming sinks.

- Pipe can be used with both of them and models stream transformations.

- Effect cannot be used with any of them and models nonstreaming components.

These components are combined using the following tools:

- for works with yield.

- (>~) works with await.

- (>->) works with both of them.

- (>>=) works with returned values.

When these monad transformers are combined, their types change to focus on the inputs and outputs that have been combined. When all inputs and outputs have been handled (i.e., they have been connected), you obtain an Effect. To stream, the last obtained Effect will be run.

The pipes package is not installed by default, so you need to install it using the following command at a terminal (if you get any error/warning message, see Chapter 26):

```
cabal install pipes
```

Let's see a simple example:

```
Prelude> import Pipes
Prelude Pipes> import qualified Pipes.Prelude as PP
Prelude Pipes PP> runEffect $ PP.stdinLn >-> PP.takeWhile
(/= "exit") >-> PP.stdoutLn
this is [--hit Enter key]
this is
a simple example [--hit Enter key]
a simple example
of using [--hit Enter key]
of using
the pipes library [--hit Enter key]
the pipes library
exit
Prelude Pipes PP>
```

Here, the first step is to import Pipes (to use runEffect) and Pipes. Prelude (to use stdinLn and takeWhile). The takeWhile function (this action works with Pipe) accepts an input as long as a predicate is satisfied (in this case, as long as the text you introduce is different from exit). The output of takeWhile becomes the input for stdinLn (this action works with Producer), which reads the string and adds a new line. To connect these two actions, you use (>->), and their result is an Effect. Finally, runEffect runs this Effect, converting it to the base monad.

Next, let's see, as an example, how stdinLn action is defined in [1].

```
import Control.Monad (unless)
import Pipes
import System.IO (isEOF)

stdinLn :: Producer String IO ()
stdinLn = do
```

```
eof <- lift isEOF          -- 'lift' an 'IO' action from the
base monad
unless eof $ do
    str <- lift getLine
    yield str              -- 'yield' the 'String'
    stdinLn                -- Loop
```

Here, the current Producer is suspended by yield, which generates a value and keeps the Producer suspended until the value is consumed. There are situations in which the value is not consumed by anybody, in this case yield will never return.

A great example of using pipes is an example of communication between a client and the server, provided in [3]. Here, the type of data that can be used in communication is defined:

```
{-# LANGUAGE DeriveGeneric #-}
module Command where
import Data.Binary
import GHC.Generics (Generic)

data Command = FirstMessage
            | DoNothing
            | DoSomething Int
            deriving (Show,Generic)

instance Binary Command
```

Next, you can see the way in which the server should handle the communication. To write the Server module, pipes-binary and pipes-network need to be installed. Open a terminal and type the following:

```
cabal install pipes-binary
cabal install pipes-network
```

Server looks like this:

```haskell
module Server where

import Pipes
import qualified Pipes.Binary as PipesBinary
import qualified Pipes.Network.TCP as PNT
import qualified Command as C
import qualified Pipes.Parse as PP
import qualified Pipes.Prelude as PipesPrelude

pageSize :: Int
pageSize = 4096

-- pure handler, to be used with PipesPrelude.map
pureHandler :: C.Command -> C.Command
pureHandler c = c  -- answers the same command that we have
receveid

-- impure handler, to be used with PipesPremude.mapM
sideffectHandler :: MonadIO m => C.Command -> m C.Command
sideffectHandler c = do
  liftIO $ putStrLn $ "received message = " ++ (show c)
  return $ C.DoSomething 0
  -- whatever incoming command 'c' from the client, answer
    DoSomething 0

main :: IO ()
main = PNT.serve (PNT.Host "127.0.0.1") "23456" $
  \(connectionSocket, remoteAddress) -> do
                putStrLn $ "Remote connection from ip = " ++
                (show remoteAddress)
                _ <- runEffect $ do
```

```
                    let bytesReceiver = PNT.fromSocket
                    connectionSocket pageSize
                    let commandDecoder = PP.parsed PipesBinary.
                    decode bytesReceiver
                    commandDecoder >-> PipesPrelude.mapM
                    sideffectHandler >-> for cat PipesBinary.
                    encode >-> PNT.toSocket connectionSocket
                    -- if we want to use the pureHandler
                    --commandDecoder >-> PipesPrelude.map
                    pureHandler >-> for cat PipesBinary.Encode
                    >-> PNT.toSocket connectionSocket
                return ()
```

Finally, the client acts like this:

```
module Client where

import Pipes
import qualified Pipes.Binary as PipesBinary
import qualified Pipes.Network.TCP as PNT
import qualified Pipes.Prelude as PipesPrelude
import qualified Pipes.Parse as PP
import qualified Command as C

pageSize :: Int
pageSize = 4096

-- pure handler, to be used with PipesPrelude.amp
pureHandler :: C.Command -> C.Command
pureHandler c = c  -- answer the same command received from the
server

-- inpure handler, to be used with PipesPremude.mapM
sideffectHandler :: MonadIO m => C.Command -> m C.Command
```

```
sideffectHandler c = do
  liftIO $ putStrLn $ "Received: " ++ (show c)
  return C.DoNothing  -- whatever is received from server,
  answer DoNothing

main :: IO ()
main = PNT.connect ("127.0.0.1") "23456" $
  \(connectionSocket, remoteAddress) -> do
    putStrLn $ "Connected to distant server ip = " ++ (show
    remoteAddress)
    sendFirstMessage connectionSocket
    _ <- runEffect $ do
      let bytesReceiver = PNT.fromSocket connectionSocket pageSize
      let commandDecoder = PP.parsed PipesBinary.decode
      bytesReceiver
      commandDecoder >-> PipesPrelude.mapM sideffectHandler >->
      for cat PipesBinary.encode >-> PNT.toSocket connectionSocket
    return ()

sendFirstMessage :: PNT.Socket -> IO ()
sendFirstMessage s = do
  _ <- runEffect $ do
    let encodedProducer = PipesBinary.encode C.FirstMessage
    encodedProducer >-> PNT.toSocket s
  return ()
```

In this example from [3], the client requests a connection through
FirstMessage. The server accepts the connection through DoSomething 0,
and then the client notices the connection is opened and sends DoNothing.
After the connection is initiated, the communication is defined through
cycles of DoSomething 0 and DoNothing. To compile, use ghc, as shown
here:

```
ghc Command.hs
ghc -main-is Client Client.hs
ghc -main-is Server Server.hs
```

Summary

In this chapter, you learned the following:

- You saw that Haskell provides great features, but they cannot be used all at once.

- You saw that, luckily, there is a library that forces the program to combine them all, namely, pipes.

- You saw a more complex example of using pipes.

References

1. Pipes.Tutorial, http://hackage.haskell.org/ package/pipes-4.3.9/docs/Pipes-Tutorial.html

2. Haskell pipes library, https://github.com/ Gabriel439/Haskell-Pipes-Library

3. Combining pipes and network communication, https://riptutorial.com/haskell/ example/29864/combining-pipes-and-network- communication

CHAPTER 20

Lens

In this chapter, you will learn about a particular type of *functional reference*. First, let's see define functional reference: *reference* means you can access and/or modify part of the values, and *functional* means that the flexibility and composability of functions are assured while accessing these parts.

Lenses are a type of functional reference, implemented in Haskell by the lens library, that represent a first-class getter and setter. With a lens, you can do the following things:

- Access a subpart

- Alter the whole by modifying a subpart

- Merge the lens with another lens to get a deeper view

When working with lenses, you need to follow some rules, depending on what you want to obtain.

- **Get-put**: If something is modified by changing just the subpart, then nothing happens.

- **Put-get**: When a particular subpart is inserted and you want to check the whole result, you will get exactly that subpart.

- **Put-put**: If subpart a is inserted, then a is modified by inserting subpart b, and this is actually the same as just inserting b.

© Stefania Loredana Nita and Marius Mihailescu 2019
S. L. Nita and M. Mihailescu, *Haskell Quick Syntax Reference*,
https://doi.org/10.1007/978-1-4842-4507-1_20

If these rules sound a bit odd, follow this chapter and things will become clearer.

The most commonly used types of lenses are as follows:

- Lens's a: When type s always contains type a, Lens s a is used to get or set the a inside of s. This is characterized as a *has-a* relationship.

- Prism's a: When type s *could* contain type a, Prism s a is used to extract a if it exists; also, given the a, it may create the s. This is characterized as an *is-a* relationship.

- Traversal's a: This finds as many a's as can be contained in s.

- Iso's a: This shows that s and a are representations of the same type.

In the previous list, the ' mark means that the lens is a simpler version of the main lens.

Let's see some simple examples of using lenses. To use them, you need to install the lens package. As usual, open a terminal and type the following:

```
cabal install lens
```

Then, in GHCi, import the library.

```
Prelude> import Control.Lens
```

The first examples are focused on tuples. The lens _1 concentrates the attention on the first element of a tuple. Some functions that can be used with _1 are view, over, and set.

```
Prelude Control.Lens> view _1 ("goal", "chaff") "goal"
```

```
Prelude Control.Lens> view _1 ("Haskell", "Lens") "Haskell"
Prelude Control.Lens> over _1 (++ " programming") ("Haskell", "Lens")
("Haskell programming","Lens")
Prelude Control.Lens> set _1 "Functional References" ("Haskell", "Lens")
("Functional References","Lens")
```

These examples are self-explanatory. Maybe over is a little more complex: the alteration is applied on the focal point _1. The three functions have an infix form: view as (^.), set as (.~), over as (%~).

The mathematical operators can be applied as lenses, in the following forms: (+~), (-~), (*~), (<>~).

By now, all these functions belong to Lens. Let's continue with Prism. The main functions in Prism are preview (^?), which can get a value from a structure, and review (#), which constructs s from a.

```
Prelude Control.Lens> preview _Left (Left "Haskell")
Just "Haskell"
Prelude Control.Lens> review _Left "Hakell"
Left "Hakell"
Prelude Control.Lens> review _Just "Hakell"
Just "Hakell"
Prelude Control.Lens> preview _Cons [1,2,3]
Just (1,[2,3])
```

Some useful functions in Traversal are traverse, which is a generalization of over, and toListOf (^..), which creates a list from what it traverses.

```
Prelude Control.Lens> (_1 . traverse) (\x -> [x, -x]) ([11,12],
"Haskell")
[([11,12],"Haskell"),([11,-12],"Haskell"),
([-11,12],"Haskell"),([-11,-12],"Haskell")]
```

```
Prelude Control.Lens> toListOf _2 (4, [1,2,3])
[[1,2,3]]
Prelude Control.Lens> toListOf _1 (4, [1,2,3])
[4]
```

Let's see a simple example of using traversals, from a great post of Chris Penner.[1] First, define a data structure like this:

```
data Transaction =
    Withdrawal {amount :: Int}
    | Deposit {amount :: Int }
  deriving Show
```

This representation is for a bank transaction, where you have the two constructors Withdrawal and Deposit, each of them with an amount value that gets the sum from either of the constructors. With a list of transactions, to focus on every element of the list using a traversal, proceed like this:

```
simpleTransactions :: Traversal' [Transaction] Transaction
simpleTransactions = traverse
```

Note that simpleTransactions has the same signature as traverse. The previous function can be successfully replaced by traverse. Moreover, it is actually recommended to use traverse instead to define your own versions.

In fact, simpleTransactions won't work. Traversal' s a may not change the structure's type or the focused value, which means with simpleTransactions, Transaction may not be changed in other type. Let's check it out with this example:

```
Prelude Control.Lens> :{
Prelude Control.Lens| someTransactions :: [Transaction]
```

---

[1]https://lens-by-example.chrispenner.ca/articles/traversals/
writing-traversals

```
Prelude Control.Lens| someTransactions = [Deposit 100,
Withdrawal 50]
Prelude Control.Lens| :}
Prelude Control.Lens> someTransactions & simpleTransactions .~
"a string"

error:
    • Couldn't match expected type 'Transaction'
                   with actual type '[Char]'
    • In the second argument of '(.~)', namely '"a string"'
      In the second argument of '(&)', namely
         'simpleTransactions .~ "a string"'
      In the expression:
         someTransactions & simpleTransactions .~ "a string"
```

The & used in lenses is defined as a flip, so it will reverse the application operator. If the traversal is allowed to change the type of the focus, it will work.

```
typeChangingTransactions :: Traversal [Transaction] [result]
Transaction result
typeChangingTransactions = traverse
```

Then you have this:

```
Prelude Control.Lens > someTransactions &
typeChangingTransactions .~ "a string"
["a string","a string"]
```

Going further with lenses, let's focus now on Iso, whose name comes from isomorphism. This means it represents a connection between equivalent types. The following is an example of Iso:

```
isoExample :: Iso' (Maybe a) (Either () a)
```

```
Prelude Control.Lens> Just "hello" ^. isoExample
Right "hello"
```

Iso has an interesting behavior because it is invertible and always succeeds, and by making some changes, you can easily get the other types of lenses. Giving up invertibility, you get a Lens; giving up successfulness you get a Prism; and giving up both, you get almost a Traversal. To actually get a Traversal, you need to go a step further and renounce having at most one target. Speaking more technically, note the following:

- The existence of i :: Iso's a says that having the value s, you also have the value a *and the inverse*. Based on this, the important functions in Iso are view i :: s -> a and review i :: a -> s. Both of them succeed and have no loss.

- The existence of l :: Lens's a says that having the value s, you have also the value a, *but the inverse way is not possible*. It is possible that the function view l :: s -> a can eliminate some information in its way, because it is not guaranteed that a conversion is lossless. So, having the a, you can't go backward to s.

- Finally, the existence of p :: Prism's a says that having the value s, *it is possible to also have the value a*, but this fact is not guaranteed. Even when converting with preview p :: s -> Maybe a is possible to fail, you still have the inverse of review p :: a -> s.

Summary

In this chapter, you learned the following:

- What lenses are
- When lenses can be used and which are the main functions for every type of lens
- How you can obtain the other lenses from Iso

Note that lenses are analogous to structures in other programming languages, such as C.

References

1. S. Fischer, H. Zhenjiang, and H. Pacheco, "A Clear Picture of Lens Laws," International Conference on Mathematics of Program Construction (Springer, 2015)

2. R. O'Connor, "Functor Is to Lens as Applicative Is to Biplate: Introducing Multiplate," arXiv preprint arXiv:1103.2841 (2011)

3. Lens: lenses, folds and traversals, `http://hackage.haskell.org/package/lens`

4. Haskell/lenses and functional references, `https://en.wikibooks.org/wiki/Haskell/Lenses_and_functional_references`

5. A little lens starter tutorial, `https://www.schoolofhaskell.com/school/to-infinity-and-beyond/pick-of-the-week/a-little-lens-starter-tutorial`

6. Writing traversals, `https://lens-by-example.chrispenner.ca/articles/traversals/writing-traversals`

7. `Control.Lens.Tutorial`, `http://hackage.haskell.org/package/lens-tutorial-1.0.3/docs/Control-Lens-Tutorial.html`

CHAPTER 21

Lazy Evaluation

You already know that Haskell is based on lazy evaluation. This means that the expressions are evaluated only when it is necessary. But what is "necessary"? In this chapter, you will get an answer to that question, and you will take a deeper look at lazy evaluation in Haskell.

First, let's take a look at strict evaluation, which is the opposite of lazy evaluation. Suppose you have this function:

```
f x y = 2*y
```

If you call `f (1234^100) 3`, in strict evaluation the first argument will be evaluated and then the second one. Looking closely at the function body, you can see that the first argument is not used. Still, it is evaluated in a strict evaluation approach, which is useless in this particular example. The advantage of strict evaluation is that it knows for sure when and in what order things will happen. For example, if you write `f(count_apples(), sing_song())` in Java, first `count_apples()` is evaluated, and then `sing_song()` is evaluated. Finally, their results are passed to `f`, which will be evaluated. When the results are not used in the body of `f`, extra work is done.

Languages are also focused on side effects. A *side effect* is any event that triggers the evaluation of an expression to interact with something outside itself. Lazy evaluation does not know when a certain expression will be evaluated, so a side effect would be useless. But a pure programming language would not make too many things; it is restrictive.

© Stefania Loredana Nita and Marius Mihailescu 2019
S. L. Nita and M. Mihailescu, *Haskell Quick Syntax Reference*,
https://doi.org/10.1007/978-1-4842-4507-1_21

In Haskell, side effects are handled in an elegant manner, through the IO monad, where they are restricted so that they do not affect the essential purity of the language.

As you saw at the beginning of the chapter, lazy evaluation means that the evaluation of an expression is postponed as much as possible; it's evaluated in the moment when it really is needed and only as far as needed, not further. The arguments of a function are just packed as unevaluated expressions called *thunks*, without doing any computation. In the example f (1234^100) 3, the arguments are packed, and the function is called immediately. As (1234^100) is not used, it remains a thunk, and no additional work is done. In other words, the expressions are evaluated when they pattern match.

Here's a simple example:

```
Prelude> pick a b c = if a > c then a else b
Prelude> pick (7+2) (9+1) (3+2)
9
```

In strict evaluation, this would be resolved as follows:

```
pick (7+2) (9+1) (3+2)
pick 9 (9+1) (3+2)
pick 9 10 (3+2)
pick 9 10 5
if 9 > 5 then 9 else 10
if True then 9 else 10
9
```

With lazy evaluation, it is resolved, beginning with the outermost expression.

```
pick (7+2) (9+1) (3+2)
if (7+2) > (3+2) then (7+2) else (9+1)
if 9 > (3+2) then 9 else (9+1)
```

```
if 9 > 5 then 9 else (9+1)
if True then 9 else (9+1)
9
```

What are the advantages of lazy evaluation? There are many, described here:

- Lazy languages are pure, which means it is difficult to identify side effects. Function reasoning is done using equality, for example, fct y = y + 5.

- In lazy languages, "value restriction" is not needed, which means that the syntax is cleaner. For example, in nonlazy languages you can use keywords like var or function to define things, but in Haskell, all these things fall into one area. Lazy languages permit you to write code in a "very functional" manner, which enables a top-down approach to coding. This feature has the advantage that things can be understood in fragments. For example, in Haskell, you can have things like this:

```
fct x y = if cond1
            then some (combinators) (applyedon
largeexpression)
            else if cond2
            then largeexpression
            else Nothing
    where some x y = ...
        largeexpression = ...
          cond1 = ...
          cond2 = ...
```

Haskell keeps the details in the where clause explicitly, because it knows that the elements in the where clause are evaluated when needed. In practice, the previous code is often written using guards.

```
fct x y
  | cond1 = some (combinators) (applyedon
largeexpression)
  | cond2 = largeexpression
  | otherwise  = Nothing
  where some x y = ...
        largeexpression = ...
        cond1 = ...
        cond2 = ...
```

- In lazy evaluation, some algorithms are expressed more elegantly. For example, in the lazy version of quicksort, the cost of looking at just the first few items is proportional to the cost of selecting them.

- Lazy evaluation lets you (re)define your own structures. In nonlazy languages, things like the following piece of code cannot be done, because both branches will be evaluated, no matter the value condition.

```
if' True x y = x
if' False x y = y
```

- Elements that deal with side effects in the type system, such as monads, work only in a lazy evaluation manner.

- Operators are short-circuited. For example, && returns false if the evaluation of the first expression is false. In this case, the second expression remains a thunk.

- It permits interesting data structures such as infinite ones. Remember the function `repeat 2` from the discussion of lists in Chapter 6.

You can find a more comprehensive description of lazy evaluation and more examples in [2].

Summary

In this chapter, you learned the following:

- What lazy evaluation is

- How lazy evaluation works

- What the advantages of lazy evaluation are

References

1. P. Hudak, J. Hughes, S. Peyton Jones, and P. Wadler, "A History of Haskell: Being Lazy with Class," in Proceedings of the Third ACM SIGPLAN Conference on History of Programming Languages, pp. 12–1 (ACM, 2007)

2. Haskell/laziness, `https://en.wikibooks.org/wiki/Haskell/Laziness`

3. Haskell/lazy evaluation, `https://wiki.haskell.org/Haskell/Lazy_evaluation`

4. T. Takenobu, "Lazy Evaluation Illustrated for Haskell Divers," `https://takenobu-hs.github.io/downloads/haskell_lazy_evaluation.pdf`

CHAPTER 22

Performance

Sometimes your program will need to meet some requirements for space and time execution. It is really important to know how the data is represented, what lazy evaluation or strict evaluation involves, and how to control the space and time behavior. In this chapter, you will learn basic techniques to improve the performance of your programs.

Type Signatures

If you don't specify the type signatures, GHC will provide you with a warning about defaults and missing types. It is important to explicitly name the types. For example, the default type for the integer values is Integer, which is ten times slower than Int.

Optimization Flags

When you are thinking about time complexity, you can use –O flags. There are several options, listed here[1]:

- No –O flag: This is the default type of compiling.

- -O0 is equivalent to the first option (no optimization at all).

---

[1]https://downloads.haskell.org/~ghc/latest/docs/html/users_guide/
using-optimisation.html#o-convenient-packages-of-optimisation-flags

© Stefania Loredana Nita and Marius Mihailescu 2019
S. L. Nita and M. Mihailescu, *Haskell Quick Syntax Reference*,
https://doi.org/10.1007/978-1-4842-4507-1_22

- -O or -O1 means, "Generate good-quality code without taking too long about it." In a terminal, you use the following command:

  ```
  ghc -c -O Program.hs
  ```

- -O2 means "Apply every nondangerous optimization, even if it means significantly longer compile times." A dangerous optimization means that you actually obtain a worse time or space complexity.

Profiling

One of the most important technique that allows you to learn about space and time allocation in Haskell is profiling.

Profiling is a technique through which you can monitor the expressions in the programs. You can observe how many times an expression runs and how much it allocates. There are three methods in which you can enable profiling: using GHC, using Stack, or using Cabal. In this section, you will use GHC's profiling.

The main steps are as follows:

1. Compile the program with the -prof option. If you want automatic annotations, use the -fprof-auto option.

2. Run the program with an option that generates the profile. For example, the option +RTS -p shows time profiling, generated in a file with the .prof extension.

3. Check the profile.

To see how it works, write the following line into a file called Main.hs:

```
main = print (fib 30)
fib n = if n < 2 then 1 else fib (n-1) + fib (n-2)
```

Save the file and then open a terminal and compile it (don't forget to change the current directory to the directory that contains the Main.hs file).

```
$ ghc -prof -fprof-auto -rtsopts Main.hs
```

The –rtsopts option enables RTS.

Next, run the program.

```
$ Main.exe +RTS -p
```

If you use Unix, type ./Main instead of Main.exe. This will print the result of fib 30 and will generate a file called Main.prof, which contains statistics about the time of execution. It looks similar to Figure 22-1.

```
Sun Mar 17 18:11 2019 Time and Allocation Profiling Report   (Final)

       Main.exe +RTS -p -RTS

    total time  =        0.46 secs   (456 ticks @ 1000 us, 1 processor)
    total alloc = 409,311,408 bytes  (excludes profiling overheads)

COST CENTRE MODULE      SRC              %time %alloc

fib         Main        Main.hs:2:1-50   100.0  100.0


                                                          individual      inherited
COST CENTRE  MODULE                      SRC              no.    entries  %time %alloc    %time %alloc

MAIN         MAIN                        <built-in>       41     0        0.0   0.0      100.0 100.0
 CAF         GHC.IO.Encoding.CodePage    <entire-module>  67     0        0.0   0.0        0.0   0.0
 CAF         GHC.IO.Encoding             <entire-module>  64     0        0.0   0.0        0.0   0.0
 CAF         GHC.IO.Handle.Text          <entire-module>  61     0        0.0   0.0        0.0   0.0
 CAF         GHC.IO.Handle.FD            <entire-module>  54     0        0.0   0.0        0.0   0.0
 CAF         Main                        <entire-module>  48     0        0.0   0.0      100.0 100.0
  main       Main                        Main.hs:1:1-21   82     1        0.0   0.0      100.0 100.0
   fib       Main                        Main.hs:2:1-50   84     2692537  100.0 100.0    100.0 100.0
  main       Main                        Main.hs:1:1-21   83     0        0.0   0.0        0.0   0.0
```

Figure 22-1. Profiling a simple program

In the first section of the file, you can see the program names and options, the total time, and the total memory used. The second section shows the costliest function (time and allocation), and the third section shows details about costs. The statistics are displayed for individual items and inherited, which includes the cost of the children of the node.

For a complete guide to profiling, consult [1]. If you want to use profiling with Cabal, consult [2]. For profiling with Stack, consult [3].

The weigh Library

The weigh library measures how much memory a value or a function uses. To install it, open a terminal and type the following:

```
cabal install weigh
```

The following is a simple example, from the GitHub page of the library[2]:

```
import Weigh

main :: IO ()
main =
  mainWith
    (do func "integers count 0" count 0
        func "integers count 1" count 1
        func "integers count 10" count 10
        func "integers count 100" count 100)
  where
    count :: Integer -> ()
    count 0 = ()
    count a = count (a - 1)
```

---

[2]https://github.com/fpco/weigh

The result will look like this:

Case	Allocated	GCs
integers count 0	16	0
integers count 1	88	0
integers count 10	736	0
integers count 100	7,216	0

Other Techniques

Here are some other techniques for obtaining good performance:

- Checking for space leaks [4]

- Setting up an isolated benchmark, using Criterion [5]

- Checking for the strictness of functions' arguments [6]

- Using a correct data structure [7]

- Checking for strictness and unpacking the types [8]

- Checking to see whether the code is polymorphic [9]

- Using the core language to generate real code before assembly (many optimizations can be done here) [10]

- Using Text or ByteString instead of String [11]

References

1. Profiling, https://downloads.haskell.org/~ghc/master/users-guide/profiling.html

2. Tutorial: profiling Cabal projects, https://nikita-volkov.github.io/profiling-cabal-projects/

3. DWARF, https://docs.haskellstack.org/en/latest/GUIDE/#dwarf

4. Detecting space leaks, http://neilmitchell.blogspot.com/2015/09/detecting-space-leaks.html?m=1

5. Criterion: a Haskell microbenchmarking library, http://www.serpentine.com/criterion/

6. Performance/strictness, https://wiki.haskell.org/Performance/Strictness

7. Specific comparisons of data structures, https://wiki.haskell.org/Performance#Specific_comparisons_of_data_structures

8. Unpacking strict fields, https://wiki.haskell.org/Performance/Data_types#Unpacking_strict_fields

9. Performance/overloading, https://wiki.haskell.org/Performance/Overloading

10. Looking at the core, https://wiki.haskell.org/Performance/GHC#Looking_at_the_Core

11. Haskell string types, http://www.alexeyshmalko.com/2015/haskell-string-types/

CHAPTER 23

Haskell Stack

Haskell Stack is a tool used to build Haskell projects and to handle its dependencies, including GHC, Cabal, a version of the Hackage repository, and a version of the Stackage package collection tool. In this chapter, you will learn the main uses of Haskell Stack.

The first step is to install Haskell Stack. On a Unix system, open a terminal and type the following:

```
curl -sSL https://get.haskellstack.org/ | sh
```

Or type the following:

```
wget -qO- https://get.haskellstack.org/ | sh
```

In Windows, go to https://get.haskellstack.org/stable/windows-x86_64-installer.exe, where you will be prompted to download the installer. Then follow the steps in the installer.

The following are the main commands for Haskell Stack:

```
stack new my-project
cd my-project
stack setup
stack build
stack exec my-project-exe
```

© Stefania Loredana Nita and Marius Mihailescu 2019
S. L. Nita and M. Mihailescu, *Haskell Quick Syntax Reference*,
https://doi.org/10.1007/978-1-4842-4507-1_23

Let's see what every command does.

- `stack new`: Creates a new directory and all the necessary files to begin a project. The structure will look like Figure 23-1.

- `stack setup`: Downloads the compiler if needed, putting it in a separate location. This means it won't make changes outside its directory.

- `stack build`: Builds a minimal project, designing reproducible builds. In this process, curated package sets are used, called *snapshots*. The main directory contains a file called `stack.yaml` representing a blueprint. It contains a reference called *resolver* that points to the snapshot used in the build process.

- `stack exec my-project-exe`: Executes a command.

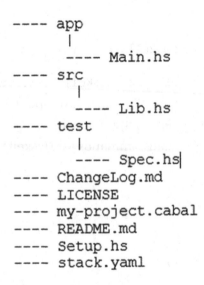

```
---- app
     |
     ---- Main.hs
---- src
     |
     ---- Lib.hs
---- test
     |
     ---- Spec.hs|
---- ChangeLog.md
---- LICENSE
---- my-project.cabal
---- README.md
---- Setup.hs
---- stack.yaml
```

Figure 23-1. The structure of a new project

Another useful command is stack install <package_name>, which installs a desired package. And, of course, don't forget about stack --help, which provides all the commands.

Now let's see a concrete example, called hello-world, inspired from [1] (note that, in this example, we will work on the Windows operating system).

Open a terminal (if you use Windows, then right-click a command prompt and choose Run as Administrator). At the terminal, choose a location for the new project.

C:\Windows\System32>cd C:\HaskellStack

Then create the project.

C:\HaskellStack>stack new hello-world new-template

This will create a new project called hello-world using the new-template template, applying an initial setup. Don't worry if you get a lot of messages. Now, you will see file in the hello-world directory, as shown in Figure 23-2.

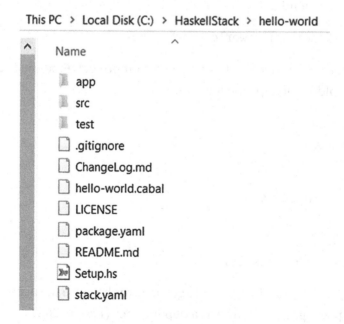

Figure 23-2. *The files in the hello-world directory*

The next step is to build the project. Before building, change the current directory to hello-world.

```
C:\HaskellStack>cd hello-world
C:\HaskellStack\hello-world>stack build
```

In this step, Stack will check for GHC and will download and install it into the global Stack root directory. This will take a while, and you will get intermediary messages about the download progress. After the build process, a library called hello-world and an executable called hello-world-exe are created in the autocreated directory .stack-work in the hello-world directory.

Further, let's run the executable.

```
C:\HaskellStack\hello-world>stack exec hello-world-exe
```

You will get a someFunc message.

In Haskell Stack, you can even test the project. To do so, use the following command:

```
C:\HaskellStack\hello-world>stack test
```

Now, let's add other functionalities to your project. Find the Lib.hs file in the src folder and type the following:

```
module Lib
    ( someFunc
    ) where

import Acme.Missiles

someFunc :: IO ()
someFunc = launchMissiles
```

If you build the project, you will get an error because the acme-missiles package is not found as a dependency (Figure 23-3).

168

```
C:\HaskellStack\hello-world>stack build
hello-world-0.1.0.0: unregistering (local file changes: src\Lib.hs)
hello-world-0.1.0.0: build (lib + exe)
Preprocessing library for hello-world-0.1.0.0..
Building library for hello-world-0.1.0.0..
[2 of 2] Compiling Lib            ( src\Lib.hs, .stack-work\dist\e626a42b\build\Lib.o )

src\Lib.hs:11:1: error:
    Could not find module `Acme.Missiles'
    Use -v to see a list of the files searched for.
   |
11 | import Acme.Missiles
   | ^^^^^^^^^^^^^^^^^^^^^

-- While building package hello-world-0.1.0.0 using:
      C:\sr\setup-exe-cache\x86_64-windows\Cabal-simple_Z6RU0evB_2.4.0.1_ghc-8.6.4.exe --builddir=.stack-work\dist\e626a
42b build lib:hello-world exe:hello-world-exe --ghc-options " -ddump-hi -ddump-to-file -fdiagnostics-color=always"
    Process exited with code: ExitFailure 1
```

Figure 23-3.   *The result of build after modifying the Lib.hs file*

To correct this, you need to modify the file package.yaml, adding the following line in the dependencies section (don't forget to save the file after editing):

- acme-missiles

If you build again, you will get another error that says it failed to construct the plan. This error is caused by the fact that the acme-missile package is not included in the long-term support (LTS) package set. To correct this, in the file stack.yaml, add the following line to create a new section called extra-deps:

extra-deps:
- acme-missiles-0.3

Now build again the project, and it will finally succeed (Figure 23-4).

```
C:\HaskellStack\hello-world>stack build
acme-missiles-0.3: download
acme-missiles-0.3: configure
acme-missiles-0.3: build
acme-missiles-0.3: copy/register
hello-world-0.1.0.0: configure (lib + exe)
Configuring hello-world-0.1.0.0...
hello-world-0.1.0.0: build (lib + exe)
Preprocessing library for hello-world-0.1.0.0..
Building library for hello-world-0.1.0.0..
[1 of 2] Compiling Lib              ( src\Lib.hs, .stack-work\dist\e626a42b\build\Lib.o )
[2 of 2] Compiling Paths_hello_world ( .stack-work\dist\e626a42b\build\autogen\Paths_hello_world.hs, .stack-work\dist\e6
26a42b\build\Paths_hello_world.o )
Preprocessing executable 'hello-world-exe' for hello-world-0.1.0.0..
Building executable 'hello-world-exe' for hello-world-0.1.0.0..
[1 of 2] Compiling Main             ( app\Main.hs, .stack-work\dist\e626a42b\build\hello-world-exe\hello-world-exe-tmp\M
ain.o ) [Lib changed]
[2 of 2] Compiling Paths_hello_world ( .stack-work\dist\e626a42b\build\hello-world-exe\autogen\Paths_hello_world.hs, .st
ack-work\dist\e626a42b\build\hello-world-exe\hello-world-exe-tmp\Paths_hello_world.o )
Linking .stack-work\dist\e626a42b\build\hello-world-exe\hello-world-exe.exe ...
hello-world-0.1.0.0: copy/register
Installing library in C:\HaskellStack\hello-world\.stack-work\install\a3f23259\lib\x86_64-windows-ghc-8.6.4\hello-world-
0.1.0.0-HAeQgBbGNEk8lRpRldzWjS
Installing executable hello-world-exe in C:\HaskellStack\hello-world\.stack-work\install\a3f23259\bin
Registering library for hello-world-0.1.0.0..
Completed 2 action(s).
```

Figure 23-4.   *Successful build*

Run the project.

```
C:\HaskellStack\hello-world>stack exec hello-world-exe
Nuclear launch detected.
```

You have created a simple project with the Haskell tool Stack.

The Haskell tool Stack is great for versioning control, focusing on reproducible build plans and multipackage projects. Of course, you can do a lot more things with it than what was presented in this chapter. For example, you can put your project into a Git repository, or you can include other projects from Git. You can find a comprehensive tutorial in [1].

Summary

In this chapter, you learned the following:

- What the Haskell tool Stack is

- What the main commands are and what they mean

- How to create, build, run, and test a new project

- How to add new dependencies to the project

References

1. User guide, `https://docs.haskellstack.org/en/stable/GUIDE/`

2. The Haskell Tool Stack, `https://docs.haskellstack.org/en/stable/README/`

CHAPTER 24

Yesod

Yesod is a web framework based on Haskell for the professional development of type-safe, REST model–based, high-performance web applications.

Yesod uses templates to generate instances for the listed entities and to produce dynamic content. The templates are based on code expression interpolations in web-like language snippets; in this way, they are fully type-checked at compile time.

Yesod divides its functionality into separate libraries. This helps you choose the functionality that you need, such as database, HTML rendering, forms, etc. You might be wondering why are we choosing Yesod here with all the web frameworks out there. There are couple of reasons why Yesod is one of the best choices.

- It's free and open source.

- It can turn runtime bugs into compile-time errors.

- Asynchronous programming is easy.

- It is scalable and performant.

- Its syntax is lightweight.

The current version of this chapter will cover version 1.6 of Yesod.

Yesod is using the Shakespearean family of template languages as a standard to HTML, CSS, and JavaScript. The templates of this language

© Stefania Loredana Nita and Marius Mihailescu 2019
S. L. Nita and M. Mihailescu, *Haskell Quick Syntax Reference*,
https://doi.org/10.1007/978-1-4842-4507-1_24

family share some common syntax and overreaching principles, such as the following:

- Well-formed content is guaranteed at compile time

- Static type safety, which helps prevent cross-site scripting (XSS) attacks

- Interpolated links that are automatically validated through type-safe URLs

Installing and Configuring Yesod

Installing Yesod can be a little tricky. Follow these steps:

1. Install and configure the Stack build tool.

   a. With your favorite browser, go to the following page: `https://haskell-lang.org/get-started/windows`.

   b. Download the Haskell Stack tool for your operating system. If your operating system is Windows, you should have a file called `stack-1.9.1-windows-x86_64-installer.exe`. Double-click and follow the steps in the installer.

   c. Once you run the file, in Windows 10 you will see the message shown in Figure 24-1. Click "More info" and click "Run anyway," as shown in Figure 24-2.

Figure 24-1. Message

Figure 24-2. *Run anyway*

d. Configure the path where Haskell Stack will be installed, as shown in Figure 24-3. We recommend you leave the default path as it is.

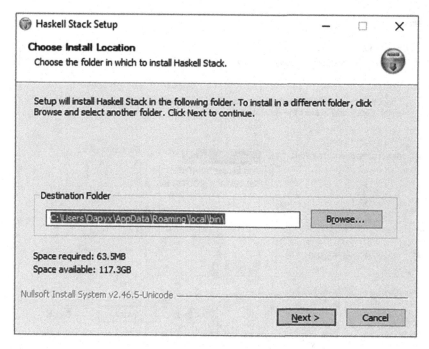

Figure 24-3. *Choosing a location*

    e. Select or deselect the components that you want or do not need. We recommend you leave everything at their defaults and click Install, as shown in Figure 24-4.

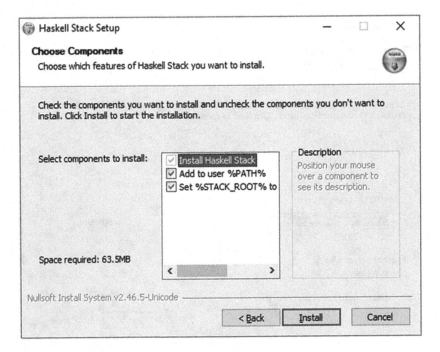

Figure 24-4. *Choosing the components*

      f. Once you see the screen in Figure 24-5, click Close.

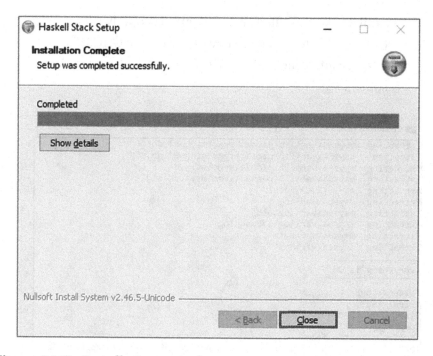

Figure 24-5. *Installation complete*

2. Copy the following code into your favorite editor (e.g., Notepad++):

```
#!/usr/bin/env stack
-- stack --install-ghc runghc

main :: IO ()
main = putStrLn "Hello World"
```

3. Save the file as HelloWorld.hs.

4. Open a terminal and run stack HelloWorld.hs.

5. If everything goes accordingly, you should see in your terminal the message "Hello World," as shown in Figure 24-6. If it is the first time you are running the program, it will take a couple of minutes

until Stack is installed and configured properly (it's possible to install the newest version of GHC underneath, which might take about 3 GB of free space).

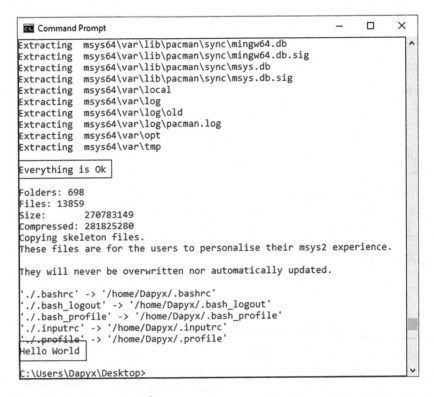

```
Command Prompt                                    —    □    ×
Extracting  msys64\var\lib\pacman\sync\mingw64.db
Extracting  msys64\var\lib\pacman\sync\mingw64.db.sig
Extracting  msys64\var\lib\pacman\sync\msys.db
Extracting  msys64\var\lib\pacman\sync\msys.db.sig
Extracting  msys64\var\local
Extracting  msys64\var\log
Extracting  msys64\var\log\old
Extracting  msys64\var\log\pacman.log
Extracting  msys64\var\opt
Extracting  msys64\var\tmp

Everything is Ok

Folders: 698
Files: 13859
Size:         270783149
Compressed: 281825280
Copying skeleton files.
These files are for the users to personalise their msys2 experience.

They will never be overwritten nor automatically updated.

'./.bashrc' -> '/home/Dapyx/.bashrc'
'./.bash_logout' -> '/home/Dapyx/.bash_logout'
'./.bash_profile' -> '/home/Dapyx/.bash_profile'
'./.inputrc' -> '/home/Dapyx/.inputrc'
'./.profile' -> '/home/Dapyx/.profile'
Hello World

C:\Users\Dapyx\Desktop>
```

Figure 24-6. *Running the program*

6. Create a new scaffold site by running the following command, shown in Figure 24-7, in your terminal: stack new my-project yesod-sqlite && cd my-project.

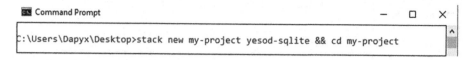

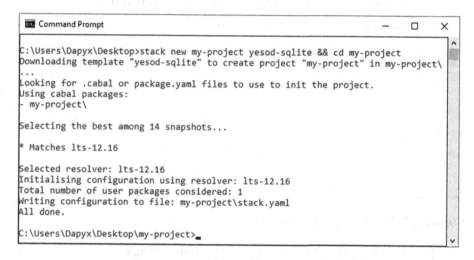

Figure 24-7. *Creating new site*

7. Once that is done, you should see the screen shown
    in Figure 24-8 in the terminal.

Figure 24-8. *Project created*

8. Install the Yesod command-line tool by running the
    command stack install yesod-bin --install-
    ghc in your terminal, as shown in Figure 24-9. So
    you don't experience any issues, it is recommended
    that you turn off your antivirus program. The firewall
    and other network settings need to be configured
    during the installation process of the Yesod
    tool. Remember also that the antivirus will run
    automatically because of some executable files that
    need to be checked.

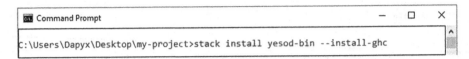

Figure 24-9. *Installing the Yesod command-line tool*

9. After a couple of minutes, if everything is OK, you should see the lines shown in Figure 24-10 in your terminal. If you see them, it means that the Yesod tool has been installed correctly.

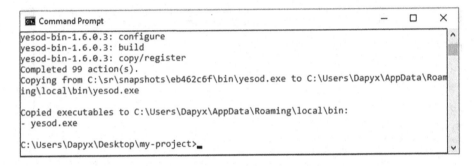

Figure 24-10. *Correctly installed*

10. Build libraries by running the command `stack build` in the terminal, as shown in Figure 24-11.

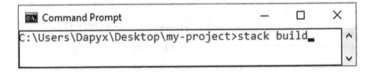

Figure 24-11. *Building libraries*

11. Launch the devel server by running the command `stack exec -- yesod devel` in the terminal, as shown in Figure 24-12. Once it starts, you will see two pop-ups with Windows Security Alert (GHC and Yesod). Click "Allow access." The process will take a couple of minutes to run.

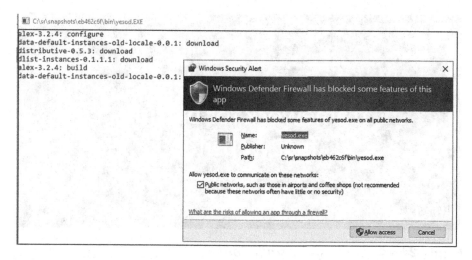

Figure 24-12. *Launching the server*

12. Once the server has been installed with success, you
     should see the screen in Figure 24-13.

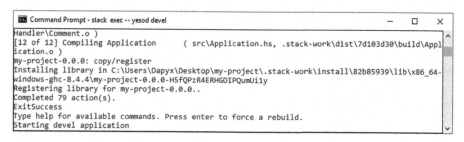

Figure 24-13. *Successful install*

13. To test Yesod, go to the browser and type http://
     localhost:3000 in the address bar. If everything
     was installed successfully, you should see the screen
     in Figure 24-14. In the terminal you will see the
     server running accordingly (Figure 24-15).

Figure 24-14. Successful installation

```
Command Prompt - stack exec -- yesod devel                          —    □    ×
Migrating: CREATE TABLE "email"("id" INTEGER PRIMARY KEY,"email" VARCHAR NOT NULL,"user_id" INTE
GER NULL REFERENCES "user","verkey" VARCHAR NULL,CONSTRAINT "unique_email" UNIQUE ("email"))
Migrating: CREATE TABLE "comment"("id" INTEGER PRIMARY KEY,"message" VARCHAR NOT NULL,"user_id"
INTEGER NULL REFERENCES "user")
Devel application launched: http://localhost:3000
Starting devel application
Devel application launched: http://localhost:3000
02/Nov/2018:23:44:56 +0200 [Debug#SQL] SELECT "id", "message", "user_id" FROM "comment" ORDER BY
"id"; []
GET /
  Accept: text/html,application/xhtml+xml,application/xml;q=0.9,*/*;q=0.8
  Status: 200 OK 0.0269836s
GET /static/css/bootstrap.css
  Params: [("etag","PDMJ7RwX")]
  Accept: text/css,*/*;q=0.1
  Status: 200 OK 0.0019841s
GET /static/tmp/autogen-8iASS6X3.css
  Accept: text/css,*/*;q=0.1
  Status: 200 OK 0.0135015s
GET /static/tmp/autogen-r3XaZuvR.js
  Accept: */*
  Status: 200 OK 0.0063868s
GET /favicon.ico
  Accept: text/html,application/xhtml+xml,application/xml;q=0.9,*/*;q=0.8
  Status: 200 OK 0s
```

Figure 24-15. Server running

14. To quit, just press Ctrl+C to stop the server. See
     Figure 24-16.

Figure 24-16. *Stopping the server*

Using Yesod in a Practical Example

This example will show some basic commands for writing a simple application in Yesod to display text on the screen. Follow these steps:

1. Open Notepad from Windows or download Notepad++ (recommended).

2. Copy and paste the following code in your editor and save the file as HelloThere.hs. As a note, the following is a general example that can be found also within other literature references.

```
{-# LANGUAGE OverloadedStrings      #-}
{-# LANGUAGE QuasiQuotes            #-}
{-# LANGUAGE TemplateHaskell        #-}
{-# LANGUAGE TypeFamilies           #-}
{-# LANGUAGE MultiParamTypeClasses #-}
import Yesod
data HelloWorld = HelloWorld
mkYesod "HelloWorld" [parseRoutes|
/ HomeR GET
|]
```

185

```
instance Yesod HelloWorld
getHomeR :: Handler Html
getHomeR = defaultLayout [whamlet|Hello
There World!|]
main :: IO ()
main = warp 3000 HelloWorld
```

3. Save the file as HelloWorld.hs in a specific path.

4. Open a command prompt and navigate to the path
    where you saved the file (e.g., in our case it will
    be D:\).

5. At the prompt, enter the command stack runghc
    HelloWorld.hs, as shown in Figure 24-17, and hit
    Enter.

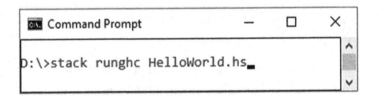

Figure 24-17. *Running the program*

6. After running the command, you should see the
    result in the command prompt window, as shown in
    Figure 24-18.

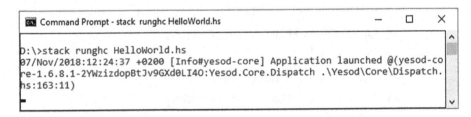

Figure 24-18. *Result of program*

7. Go to the browser and open a new tab. Enter
   the following address: http://localhost:3000.
   If everything is OK, you should see on the screen
   the result, as shown in Figure 24-19.

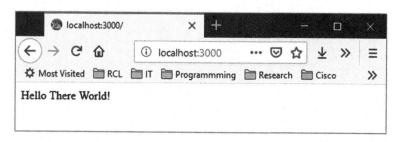

Figure 24-19. *http://localhost:3000*

Summary

In this chapter, you learned about the Yesod web framework and saw how
to use it to start developing applications.

References

1. Yesod (web framework), https://en.wikipedia.
   org/wiki/Yesod_(web_framework)

2. Yesod, https://github.com/yesodweb/yesod

3. Yesod: creation of type-safe, RESTful web
   application, http://hackage.haskell.org/
   package/yesod

4. Yesod, web framework, https://www.yesodweb.
   com/book

CHAPTER 25

Haskell Libraries

In programming, libraries are collections of precompiled routines that can be used in other programs. Usually, a library stores frequently used routines (for example, reading or printing an input), or it is specialized for a particular subject (for example, the libraries used in the statistics field). In this chapter, you will learn about the main libraries in Haskell and when you can use them to create powerful programs.

Prelude

The main library in Haskell is Prelude, which is imported by default when you install GHC.

Prelude contains standard types and classes, operations on lists, conversion techniques from a value to a String and vice versa, basic input and output operations and functions, basic exception handling, and functions for types.

Here is an example that handles exceptions using error:

```
divide :: Float -> Float -> Float
divide x 0 = error "Cannot divide by 0"
divide x y = x / y
```

© Stefania Loredana Nita and Marius Mihailescu 2019
S. L. Nita and M. Mihailescu, *Haskell Quick Syntax Reference*,
https://doi.org/10.1007/978-1-4842-4507-1_25

After loading and compiling, let's test it.

```
Prelude> :load Division.hs
[1 of 1] Compiling Main                 ( Division.hs, interpreted )
Ok, one module loaded.
*Main> divide 3 5
0.6
*Main> divide 10 0
*** Exception: Cannot divide by 0
CallStack (from HasCallStack):
  error, called at Division.hs:2:14 in main:Main
```

The previous example is simple. If the second argument of the division function is 0, then you print an error message; otherwise, you print the result of division between the two arguments.

Haskell 2010 Libraries

The language and library specification of Haskell 2010 contains a collection of libraries with basic functionalities included in all Haskell implementations. The libraries are maintained by the Haskell process, and you can find the complete list of Haskell 2010 libraries in [1].

Two important packages are base (which includes modules like Control.Monad, Data.List, and System.IO) and vector (which includes modules such as Data.Vector).

Because you have already worked with elements from base, this example involves vectors. To use functions from vector, you need to install it with the following command at a terminal:

```
cabal install vector
```

You can create a vector from a list, as shown here:

```
Prelude> import Data.Vector
Prelude Data.Vector> let a = fromList [5,6,7]
Prelude Data.Vector> a
[5,6,7]
Prelude Data.Vector> :t a
a :: Num a => Vector a
```

Here you can see which element has index 1 (note the indexing begins from 0):

```
Prelude Data.Vector> a ! 1
6
```

GHC Bootstrap Libraries

GHC Bootstrap libraries are an extension of the Haskell 2010 libraries, used to build GHC itself. Examples of such libraries are haskeline and integer-gmp., For this chapter, you don't need packages from this collection.

Core Libraries and Haskell Platform Libraries

Core libraries are part of the management process, defining the basic APIs that can be used in any Haskell implementation, the packages for backward compatibility, or the packages that are needed to link things in the Haskell platform together. Examples of such libraries and packages are the Monad transformer library, random, and parallel.

The following is an example of using the random library, where randomRs generates a random sequence:

```
import System.Random

main = do
  g <- getStdGen
  print $ take 10 (randomRs ('a', 'z') g)
```

In addition to the core libraries, the Haskell platform libraries contain more complex packages such as attoparsec, network, and QuickCheck.

The Hackage Database

The Hackage database contains a huge list of libraries, specializing in a large range of subjects, such as blockchain, chemistry, files, hydrology, scientific simulation, and so on. You can find the complete list of packages grouped by subjects in [2].

Summary

In this chapter, you learned about classified packages and libraries in Haskell. Note that because Haskell is an open source platform, anyone can create libraries. A short tutorial of the processes involved in creating and submitting a library can be found in [3] or [4].

References

1. Part II, the Haskell 2010 libraries, `https://www.haskell.org/onlinereport/haskell2010/haskellpa2.html`

2. Packages by category, `http://hackage.haskell.org/packages/`

3. How to write a Haskell program, `https://wiki.haskell.org/How_to_write_a_Haskell_program`

4. Creating a package, `https://downloads.haskell.org/~ghc/7.0.4/docs/html/Cabal/authors.html`

5. Applications and libraries, `https://wiki.haskell.org/Applications_and_libraries`

CHAPTER 26

Cabal

Haskell includes a standard package system called Cabal, which is used to configure, build, install, and (re)distribute Haskell software. In this chapter, you will learn how to use Cabal to increase your productivity.

Cabal is that part of Haskell that helps you to manage packages. It draws the packages from Hackage, the archive of Haskell that contains a large number of libraries in the Cabal package format.

Packages help developers to distribute and (re)use software. A package system is an important component because it centralizes reusable software that can be shared by many developers.

The components of Cabal are called *Cabal packages* and can be independent, but they can also depend on one another. Cabal knows this, and when a package is installed, it also installs all the dependencies of that package. Note that packages are not part of the Haskell language (because they are not included by default in the Haskell installation), but you can think of them as features resulting from a combination of Cabal and GHC.

The command that is used to build and install a package is `cabal`. Let's see a simple example. Open a terminal (if you work on Windows OS, right-click the terminal and choose Run as Administrator). Type `cabal --help`. This command will provide a list of options for `cabal`, as shown in Figure 26-1.

© Stefania Loredana Nita and Marius Mihailescu 2019
S. L. Nita and M. Mihailescu, *Haskell Quick Syntax Reference*,
https://doi.org/10.1007/978-1-4842-4507-1_26

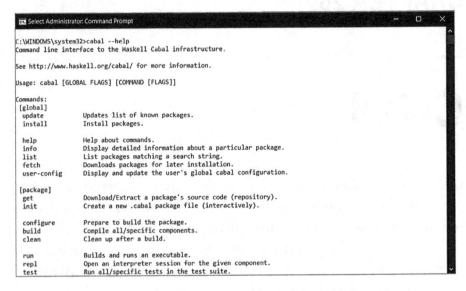

Figure 26-1. *The result of the cabal --help command*

A Cabal package contains the following items:

- The software, i.e., libraries, executable, tests

- The .cabal file, which is a metadata file with information about that package

- The Setup.hs file, which is a standard interface, based on which the package is built

Many developers use cabal to install a package. To install a package, all you need to do is to check its page at http://hackage.haskell.org and then type at the terminal cabal install, followed by the package name. For example, let's install the errors package (this package helps to handle errors), as shown in Figure 26-2.

Figure 26-2. *Installing the errors package*

If you already have installed a desired package, you will receive the message shown in Figure 26-3.

```
Administrator: Command Prompt
C:\WINDOWS\system32>cabal install exceptions
Resolving dependencies...
All the requested packages are already installed:
exceptions-0.10.0
Use --reinstall if you want to reinstall anyway.

C:\WINDOWS\system32>
```

Figure 26-3. *Message for an already installed package*

If you feel everything is OK, you don't need to reinstall the package.

Let's see some warnings. If you received a warning message when you tried to install the pipes package in Chapter 19, you are in the right place. Let's try to install the pipes package; you will receive the message shown in Figure 26-4.

```
C:\WINDOWS\system32>cabal install pipes
Warning: The package list for 'hackage.haskell.org' does not exist. Run 'cabal
update' to download it.
cabal: There is no package named 'pipes'.
You may need to run 'cabal update' to get the latest list of available
packages.

C:\WINDOWS\system32>
```

Figure 26-4. *A warning message*

197

This means your package archive is out-of-date, but don't worry—all you need to do is to follow the steps in the message. So, update the archive, as shown in Figure 26-5.

Figure 26-5. Updating the archive of packages

This may take a few minutes. Next, type the installation command, as shown in Figure 26-6.

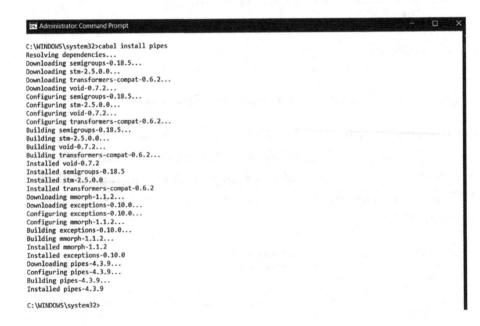

Figure 26-6. Successful installation of a package

That's it, you just installed the pipes package.

Note that in Figure 26-6 that we built and installed all the packages that pipes needs to work properly.

In addition, these steps work for all packages for which you receive a similar warning message, not just for pipes. Now that you know what happens in the terminal, we will show you just the commands.

You can also install a package from a local source (the archive for a package usually ends with .tar.gz), as shown here:

```
$ cabal install pipes-4.3.9.tar.gz
Resolving dependencies...
In order, the following will be installed:
pipes-4.3.9 (reinstall)
Warning: Note that reinstalls are always dangerous. Continuing
anyway...
Configuring pipes-4.3.9...
Building pipes-4.3.9...
Installed pipes-4.3.9
```

Note   When you install a package from a local source, the current directory from the terminal needs to be the same as the local path where the package is located.

Cabal is not used just to install packages. It can do a lot more, such as the following:

- Build a software system

- Configure a system

- Package a system for distribution

- Run tests

- Generate documentation

- Automatically manage the packages

- Archive online and local packages in Cabal form

For example, here are the commands used to build and install a system package:

```
$ runhaskell Setup.hs configure --ghc
$ runhaskell Setup.hs build
$ runhaskell Setup.hs install
```

On the first line, the system is prepared to build the software using GHC, on the second line it is actually built, and on the third line it installs the package by copying the build results to a permanent place and registering the package with GHC.

There are lot things that can be done with Cabal that require more advanced programming skills than shown here. You can find a complete guide to using Cabal on the official page.[1]

Summary

In this chapter, you learned the following:

- What Cabal is and what it can do

- What the structure of a package is

- How to install packages and resolve some error or warning messages

---

[1]https://www.haskell.org/cabal/users-guide/

References

1. I. Jones, "The Haskell Cabal: A Common
    Architecture for Building Applications and
    Libraries" (2005)

2. P. Hudak, "A History of Haskell: Being Lazy with
    Class" in proceedings of the Third ACM SIGPLAN
    Conference on History of Programming Languages
    (ACM, 2007)

3. Cabal user guide, https://www.haskell.org/
    cabal/users-guide

Index

© Stefania Loredana Nita and Marius Mihailescu 2019
S. L. Nita and M. Mihailescu, *Haskell Quick Syntax Reference*,
https://doi.org/10.1007/978-1-4842-4507-1

Printed in the United States
by Bookmasters

Printed in the United States
By Bookmasters